Systems Thinking 2.0

-- Architectural Thinking Using the SBC Architecture Description Language --

William S. Chao

2

Structure-Behavior Coalescence

$$\text{Systems Architecture} = \text{Systems Structure} + \text{Systems Behavior}$$

CONTENTS

6

PREFACE

Human beings have employed the notion of systems so widely in all kinds of scientific studies. Systems definition is an artifact created by humans to describe what a system is. A system has been defined, by systems thinking 1.0, hopefully to be an integrated whole, embodied in its assembled components, their interrelationships with each other and the environment. This systems thinking 1.0 defining a system possesses one cardinal deficiency. The deficiency comes from that it does not describe the integration of systems structure and systems behavior.

Systems structure and systems behavior are the two most significant views of a system. In order to achieve a truly integrated whole of a system, we first need to integrate the systems structure and systems behavior together. In other words, integration of systems structure and systems behavior results in the integration of a whole system. Since systems thinking 1.0 does not describe the integration of systems structure and systems behavior, very likely it only hopes and will never be able to really form an integrated whole of a system. In this situation, systems thinking 1.0 is powerless in defining a system appropriately.

Structure-behavior coalescence (SBC) provides an elegant way to integrate the systems structure and systems behavior of a system. A system is therefore redefined, by systems thinking 2.0 (architectural thinking), truly to be an integrated whole, through structure-behavior coalescence, embodied in its assembled components, their interactions (or handshakes) with each other and the environment. Systems thinking 2.0 uses the SBC architecture description language (SBC-ADL) to formally define the essence of a system and its details at the same time. SBC-ADL contains six fundamental diagrams: a) architecture hierarchy diagram, b) framework diagram, c) component operation diagram, d) component connection diagram, e) structure-behavior coalescence diagram and f) interaction flow diagram. Since systems thinking 2.0 demands the integration of systems structure and systems behavior, definitely it is able to form an integrated whole of a system. In this situation, systems thinking 2.0 is fully capable of defining a system.

In this book, we shall dwell on systems thinking 2.0 through the application of SBC-ADL. By this book's introduction and elaboration of SBC-ADL, all readers will understand clearly how systems thinking 2.0 helps us define a truly integrated whole of a system.

ABOUT THE AUTHOR

Dr. William S. Chao is the CEO & founder of SBC Architecture International®. SBC (Structure-Behavior Coalescence) architecture is a systems architecture which demands the integration of systems structure and systems behavior of a system. SBC architecture applies to hardware architecture, software architecture, enterprise architecture, knowledge architecture and thinking architecture. The core theme of SBC architecture is: "Architecture = Structure + Behavior."

William S. Chao received his bachelor degree (1976) in telecommunication engineering and master degree (1981) in information engineering, both from the National Chiao-Tung University, Taiwan. From 1976 till 1983, he worked as an engineer at Chung-Hwa Telecommunication Company, Taiwan.

William S. Chao received his master degree (1985) in information science and Ph.D. degree (1988) in information science, both from the University of Alabama at Birmingham, USA. From 1988 till 1991, he worked as a computer scientist at GE Research and Development Center, Schenectady, New York, USA.

Dr. William S. Chao has been teaching at National Sun Yat-Sen University, Taiwan since 1992 and now serves as the president of Association of Enterprise Architects, Taiwan Chapter. His research covers: systems architecture, hardware architecture, software architecture, enterprise architecture, knowledge architecture and thinking architecture.

PART I: SYSTEMS THINKING 1.0 VERSUS SYSTEMS THINKING 2.0

14

Chapter 1: Introduction to Systems

The word "system" originates from the Greek term, systēma, meaning "composition" or "whole". The notion of systems has been so widely used in all kinds of scientific studies such as systems analysis and design [Hoff10, Shel11], systems architecting [Maie09, Mull11], systems architecture [Burd10, Roza11], systems bible [Gall03, Kill09], systems biology [Klip09, Voit12], system dynamics [Ogat03, Palm09], systems ecology [Jorg12, Odum94], systems engineering [Beam90, Kass07, Koss11], systems medicine [Pork78, Weil04, Weil00], systems modeling [Frie11], systems physiology [Raff11, Sher09], systems requirement [Bere09, Grad06], systems science [Warf06], systems theory [Bert69, Luhm12], systems thinking [Chec99, Ghar11, Mead08], systems view [Bert81, Lasz96].

In this chapter, we first introduce the systems thinking 1.0. We then introduce the physical and virtual systems. A physical system exists in the physical, concrete, or real world. A virtual system exists in the virtual, abstract, or notional world. The final section of this chapter will introduce the boundary and environment of a system. A system has a boundary. The system itself is inside the boundary and the environment is outside the boundary.

1-1 Systems Thinking 1.0

All things that amaze us as something independent are essentially parts of a system. We usually call the parts of a system its components. Every system is something the whole. Systems emphasize the holistic vision.

The need for defining a system arises because any real-life system is inherently complicated. It is impossible to comprehend fully the intricate interrelationships of any system of the real world with its environment, or to describe all its components and each of its details. Systems definition is an artifact created by humans to describe what a system is [Kapo94].

Systems thinking 1.0 defines a system, in Figure 1-1, hopefully to be an integrated whole, embodied in its assembled components, their interrelationships with each other and the environment [Chec99, Ghar11, Mead08].

> A system, hopefully is an integrated whole,
> embodied in its assembled components,
> their interrelationships with each other and the environment.

Figure 1-1   Systems Thinking 1.0 Defining a System

Components are sometimes labeled as parts, entities, objects, building blocks and non-aggregated systems [Chao09, Chao14]. Interrelated components make a system not only a whole but also hopefully an integrated whole.

A system defined by systems thinking 1.0 has the following characteristics: 1) hopefully, it is an integrated whole; 2) it is embodied in its assembled components; 3) components are interrelated with each other and the environment; and 4) it uses structural decomposition [Chao12, Ghar11] rather than functional decomposition [Scho10].

The structural decomposition method is to decompose a system into a number of components, as shown in Figure 1-2. Breaking down a large problem into a number of components to solve, is a relatively preferred method.

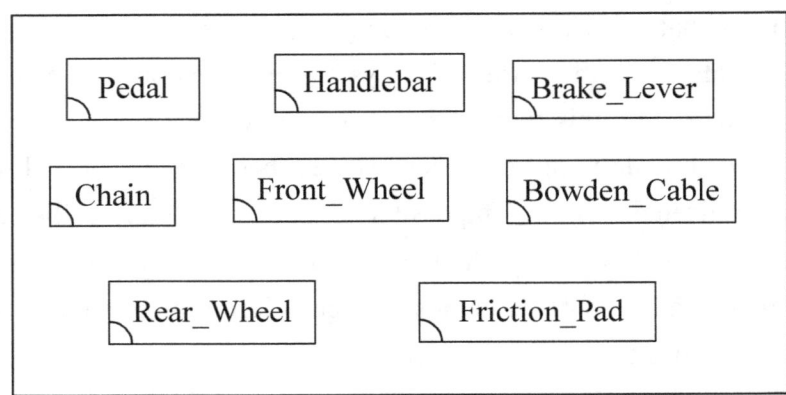

Figure 1-2   Structural Decomposition Method

The functional decomposition method is to decompose a system into a number of functions, as shown in Figure 1-3. Breaking down a large problem into a number of functions to solve, is a relatively non-preferred method.

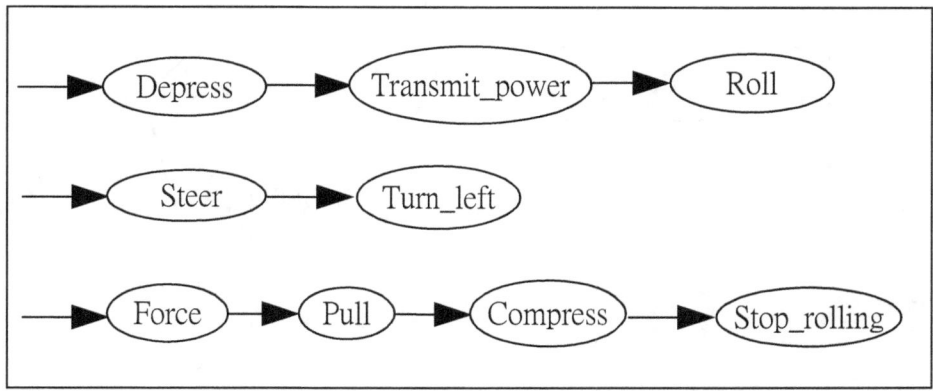

Figure 1-3   Functional Decomposition Method

Systems definition is an artifact created by humans to describe what a system is. Without a systems definition, everybody has his own saying about a system and never be able to reach a consensus. For example, Bruce Kennedy thinks the *Classroom_4069* is embodied in its assembled components of *desk_1* and *chair_1*, their interrelationships with each other and the environment; Tom Johnson thinks the *Classroom_4069* is embodied in its assembled components of *desk_1*, *chair_1*, *chair_2* and *chair_3*, their interrelationships with each other and the environment. It is impossible for Bruce Kennedy and Tom Johnson to work together on the *Classroom_4069* if they can not reach a common definition.

To resolve the conflict between Bruce Kennedy and Tom Johnson, here comes systems thinking 1.0 defining the *Classroom_4069*, shown in Figure 1-4, hopefully to be an integrated whole, embodied in its assembled components of *desk_1*, *chair_1* and *chair_2*, their interrelationships with each other and the environment.

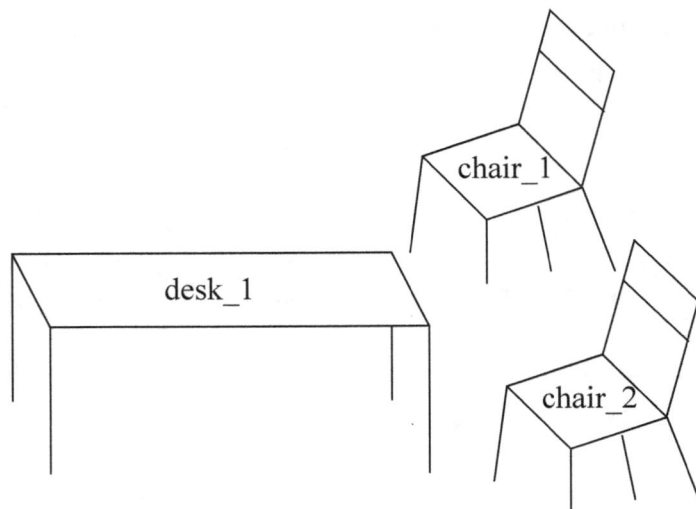

Figure 1-4   Systems Thinking 1.0 Defining the *Classroom_4069*

As a second example, systems thinking 1.0 defines a *car*, shown in Figure 1-5, hopefully to be an integrated whole, embodied in its assembled components of *body* and *wheels*, their interrelationships with each other and the environment.

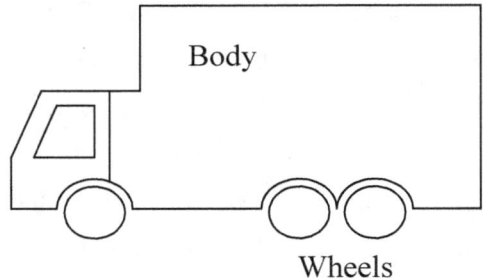

Figure 1-5   Systems Thinking 1.0 Defining a *Car*

As a third example, systems thinking 1.0 defines a *sandwich*, shown in Figure 1-6, hopefully to be an integrated whole, embodied in its assembled components of *fillings* and *slice_of_bread*, their interrelationships with each other and the environment.

Figure 1-6　Systems Thinking 1.0 Defining a *Sandwich*

1-2 Physical and Virtual Systems

In general, systems are divided into two categories: 1) physical systems and 2) virtual systems.

A physical system exists in the physical world [Acko68]. A physical system is also called a concrete or real system. For example, a *telephone* composed of *microphone, earphone and keypad*, shown in Figure 1-7, is a physical, concrete, or real system.

Figure 1-7　A Telephone is a Physical System

As a second example, a *stool* composed of *seat* and *legs*, shown in Figure 1-8, is a physical, concrete, or real system.

Figure 1-8   A Stool is a Physical System

A virtual system is a system that is composed of non-physical components, i.e., ideas, thoughts, or notions. A virtual system exists in the virtual, abstract, or notional world. For example, the "*Snow White and the Seven Dwarfs*" fairy tale composed of "*Snow White*" and "*Seven Dwarfs*," shown in Figure 1-9, is a virtual, abstract, or notional system.

Figure 1-9   *Snow White and the Seven Dwarfs* is a Virtual System

As a second example, the "*Multi-Tier Personal Data System*" software composed of *MTPDS_GUI*, *Age_Logic*, *Overweight_Logic* and *Personal_Database*, shown in Figure 1-10, is a virtual, abstract, or notional system.

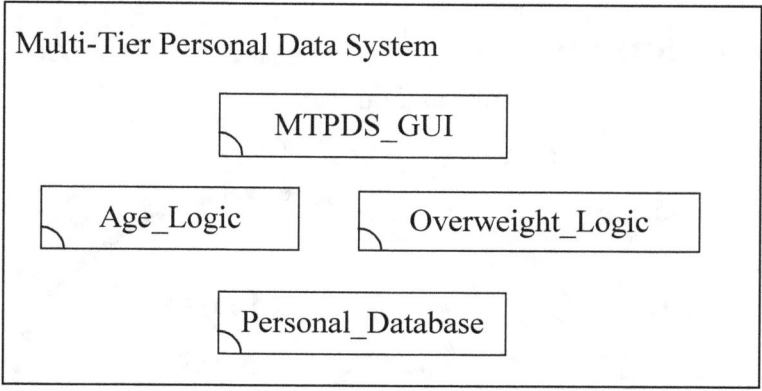

Figure 1-10   *Multi-Tier Personal Data System* is a Virtual System

1-3 Boundary and Environment of a System

We scope a system by defining its boundary as shown in Figure 1-11. All components of the system are inside the boundary while the environment is outside the boundary.

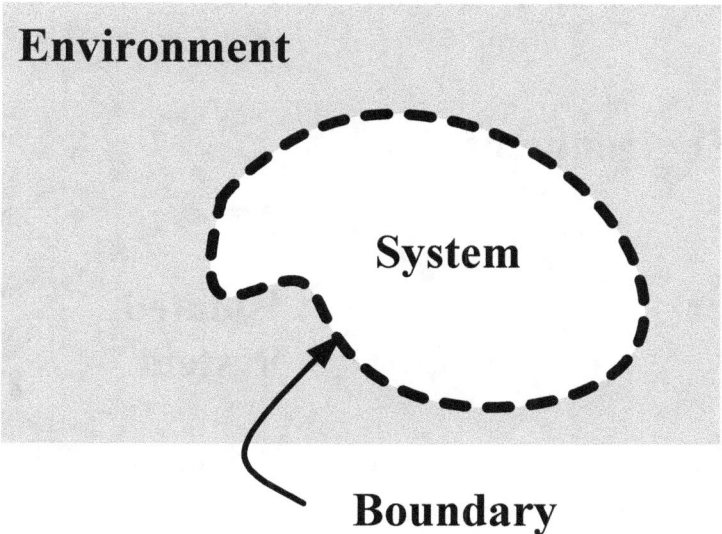

Figure 1-11   Boundary and Environment of a System

The environment is also known as the surroundings. A system may or may not interrelate with the environment. An open system interrelates with the environment through the exchange of matter, energy, data, information, or message as shown in Figure 1-12.

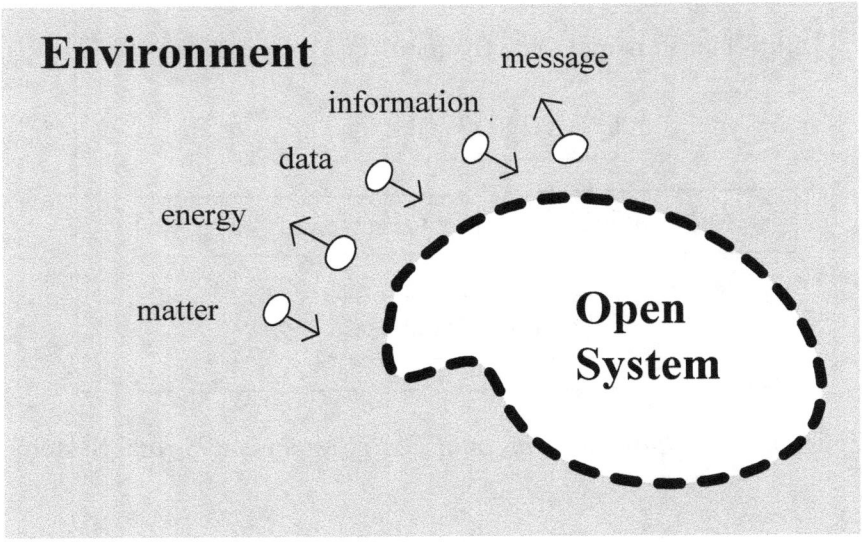

Figure 1-12   Open System Interrelates with the Environment

An isolated system does not interrelate with the environment at all. There is no exchange of matter, energy, data, information, or message between the isolated system and the environment as shown in Figure 1-13.

Figure 1-13   Isolated System Does Not Interrelate with the Environment

Chapter 2: Systems Structure and Systems Behavior

Systems structure and systems behavior are the two most significant views of a system. Systems structure, defined by components, their operations and their composition, refers to the type of connection between the components of a system. Systems behavior, defined by the interrelationships between and among the components and environment, refers to the interconnectivities a system in conjunction with its environment.

2-1 Structure of Systems

Every system forms a whole. In general, structure of systems is the type of connection between the components of a system. More specifically, we define the structure of a system by 1) components, 2) their operations and 3) their composition.

Components are something relatively indivisible in one system [Hoff10, Shel11]. For example, *Head*, *Hands* and *Feet* are components of a *robot* system as shown in Figure 2-1.

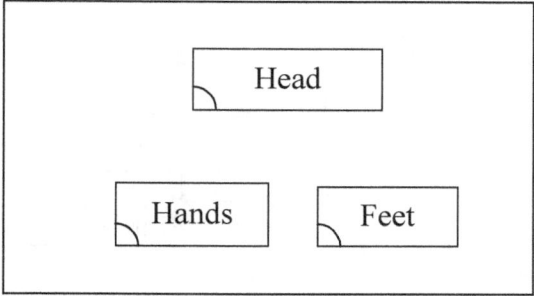

Figure 2-1   Components of a *Robot* System

An operation provided by each component represents a procedure or method or function of the component [Chao09, Chao12, Chao14]. Each component in a system must possess at least one operation. Figure 2-2 shows the operations of all components of a *robot* system. In the figure, component *Head* has two operations: *Receive_Write_Signal* and *Receive_Walk_Signal*; component *Hands* has one operation: *Move_Hand*; component *Feet* has one operation: *Move_Foot*.

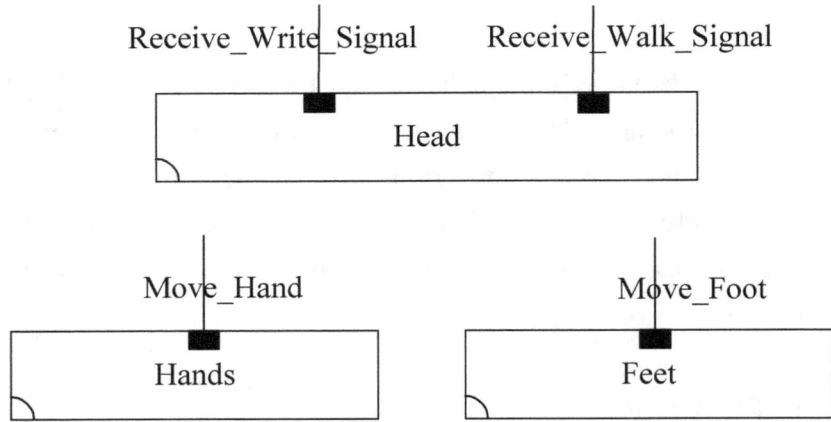

Figure 2-2   Operations of all Components of a *Robot* System

Composition of components defines the structural composition and decomposition of a system. For example, Figure 2-3 shows that in a *robot* system *Robot* is structurally composed of *Head* and *Limb*; *Limb* is structurally composed of *Hands* and *Feet*.

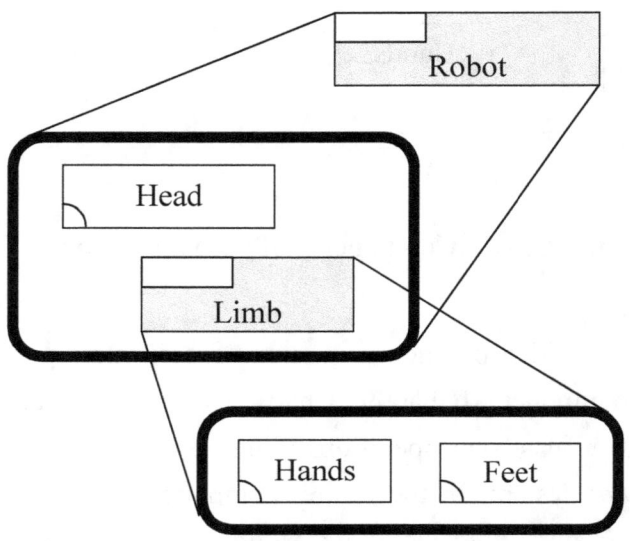

Figure 2-3   Structural Composition of a *Robot* System

2-2 Behavior of Systems

Systems behavior refers to the interrelationships a system in conjunction with its environment. It is the response of a system to various stimuli, whether internal or external, conscious or subconscious, overt or covert, and voluntary or involuntary.

For example, Figure 2-4 demonstrates two individual behaviors: *Writing* and *Walking* that refers to the interrelationships a *robot* system in conjunction with its environment.

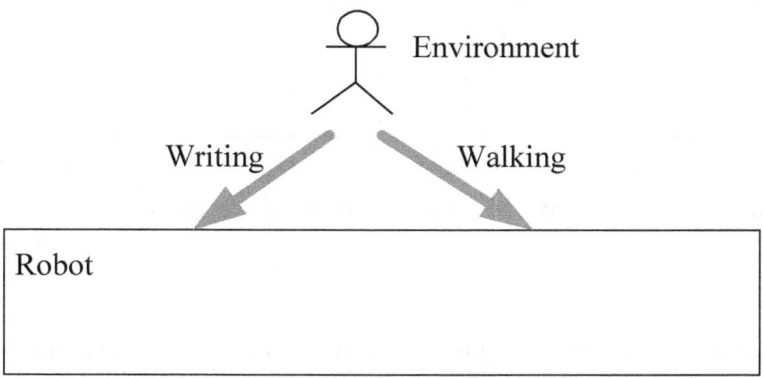

Figure 2-4   Behaviors of a *Robot* System

For each behavior, the environment always initiates the interrelationship and will lead more follow-up interrelationships to be realized among components. For example, Figure 2-5 demonstrates that interrelationships between and among the environment and the *Head, Hand*s components shall draw forth the *Writing* behavior.

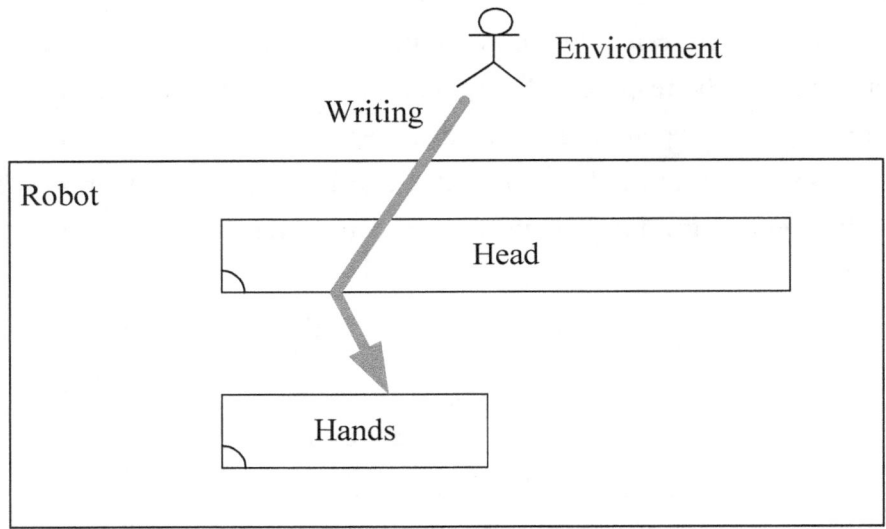

Figure 2-5   Interrelationships that Draw forth the *Writing* Behavior.

As a second example, Figure 2-6 demonstrates that interrelationships among the environment and the *Head*, *Feet* components shall draw forth the *Walking* behavior.

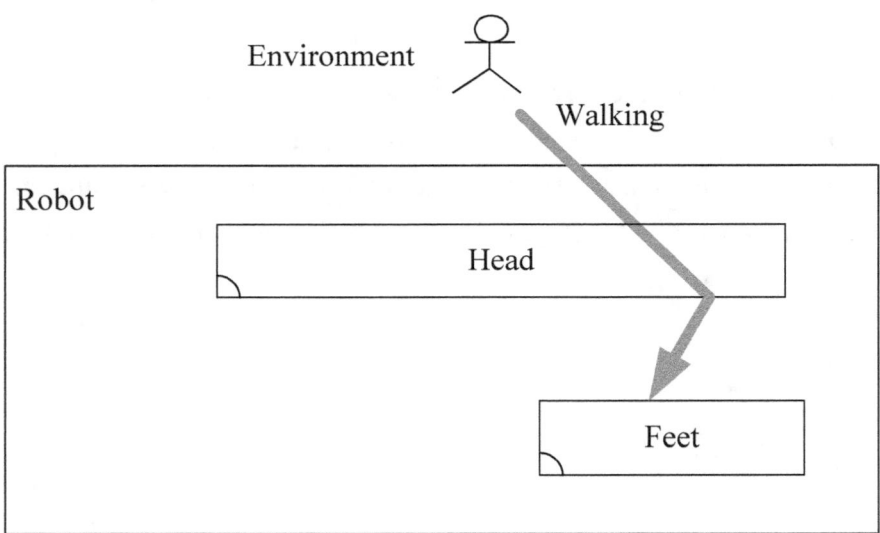

Figure 2-6   Interrelationships that Draw forth the *Walking* Behavior.

Chapter 3: Structure-Behavior Coalescence

A system has been defined hopefully to be an integrated whole, embodied in its assembled components, their interrelationships with each other and the environment. Since systems structure and systems behavior are the two most prominent views of a system, integrating the systems structure and systems behavior apparently is the best way to achieve a truly integrated whole of a system. Because systems thinking 1.0 does not describe the integration of systems structure and systems behavior, very likely it will never be able to actually form an integrated whole of a system.

Structure-behavior coalescence (SBC) provides an elegant way to integrate the systems structure and systems behavior, and hence achieves a truly integrated whole, of a system. A truly integrated whole sets a path to achieve the desired systems definition. SBC facilitates an integrated whole. Therefore, we conclude that SBC sets a path to achieve the systems definition. Systems thinking 2.0 (architectural thinking) uses the SBC approach and is highly adequate in defining a system.

3-1 Integrated Whole to Achieve the Systems Definition

A system has been defined hopefully to be an integrated whole, embodied in its assembled components, their interrelationships with each other and the environment. In other words, an integrated whole sets a path to achieve the systems definition as shown in Figure 3-1.

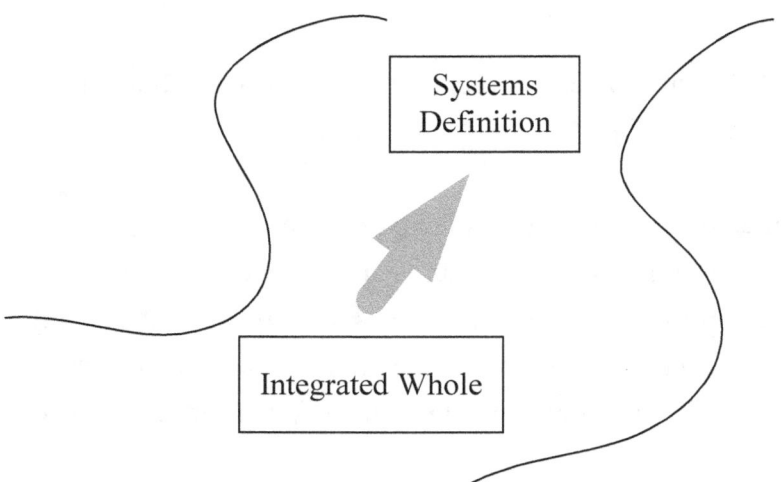

Figure 3-1　　Integrated Whole to Achieve the Systems Definition

In one systems definition, different systems structures may draw forth the same integrated whole as shown in Figure 3-2.

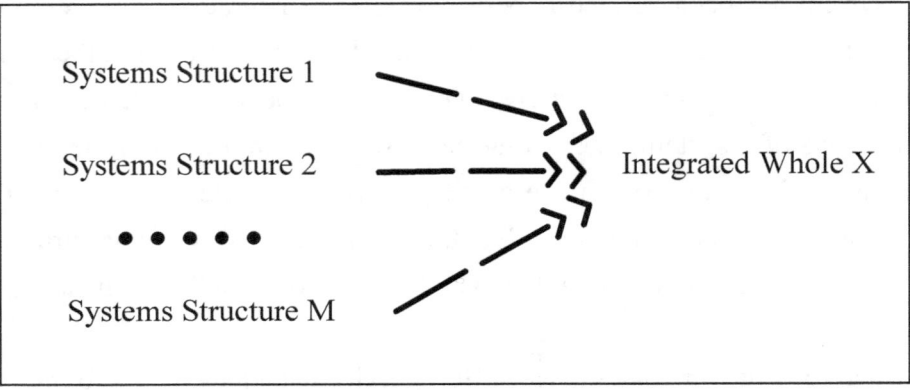

Figure 3-2   Different Systems Structures Draw Forth the Same Integrated Whole

Since there is only one systems structure exists in one systems definition, one systems structure will draw forth one integrated whole as shown in Figure 3-3.

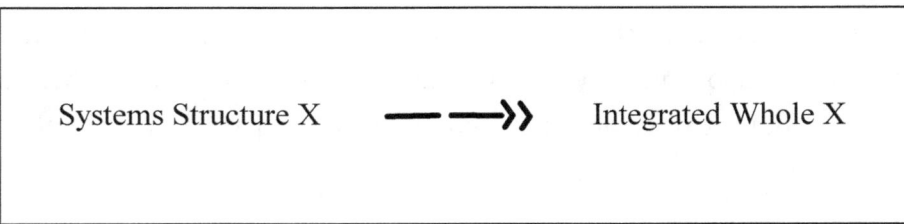

Figure 3-3   One Systems Structure Draws Forth One Integrated Whole

We conclude that in one systems definition, an integrated whole must be attached to or built on a systems structure. In other words, an integrated whole shall not exist alone; it must be loaded on a systems structure just like a cargo is loaded on a ship as shown in Figure 3-4. There will be no integrated whole if there is no systems structure. A stand-alone integrated whole with no systems structure is not meaningful.

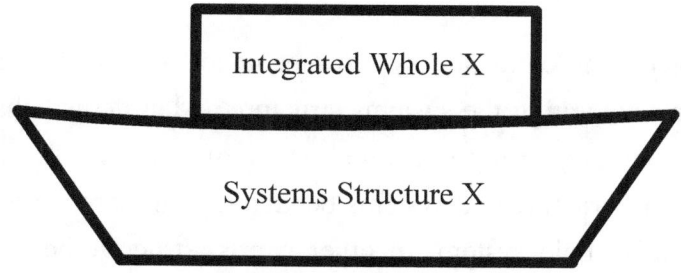

Figure 3-4   An Integrated Whole Must be Loaded on a Systems Structure

3-2 Integrating the Systems Structures and Systems Behaviors

By integrating the systems structures and systems behaviors, we obtain structure-behavior coalescence (SBC) within a system. Since systems structures and systems behaviors are so tightly integrated, we sometimes claim that the core theme of structure-behavior coalescence is: "Systems Architecture = Systems Structure + Systems Behavior," as shown in Figure 3-5.

Figure 3-5   Core Theme of Structure-Behavior Coalescence

So far, integrating the systems structure and systems behavior has never been proposed or suggested besides the SBC approach. In most cases, systems behaviors are separated from systems structures when defining a system [Hoff10, Pres09, Shel11, Somm06].

3-3 Structure-Behavior Coalescence to Facilitate an Integrated Whole

Since systems structure and systems behavior are the two most prominent views of a system, integrating the systems structure and systems behavior apparently is the best way to achieve a truly integrated whole of a system. If we are not able to integrate the systems structure and systems behavior, then there is no way that we are able to integrate the whole system. In other words, structure-behavior coalescence (SBC) facilitates a truly integrated whole as shown in Figure 3-6.

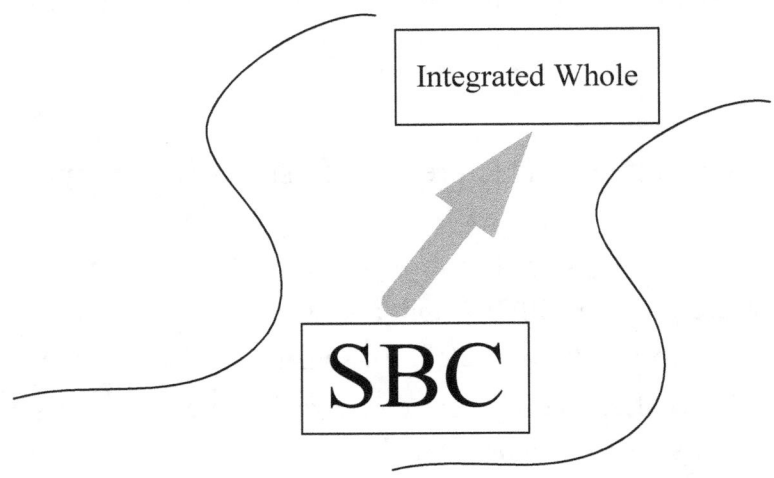

Figure 3-6   SBC Facilitates an Integrated Whole

Since systems thinking 1.0 does not describe the integration of systems structure and systems behavior, very likely it will never be able to actually form an integrated whole of a system. In this situation, systems thinking 1.0 is powerless in defining a system adequately.

3-4 Structure-Behavior Coalescence to Achieve the Systems Definition

Figure 3-1 declares that an integrated whole sets a path to achieve the desired systems definition. Figure 3-6 declares that structure-behavior coalescence facilitates a truly integrated whole.

Combining the above two declarations, we conclude that the structure-behavior coalescence (SBC) approach sets a path to achieve the systems definition as shown in Figure 3-7.

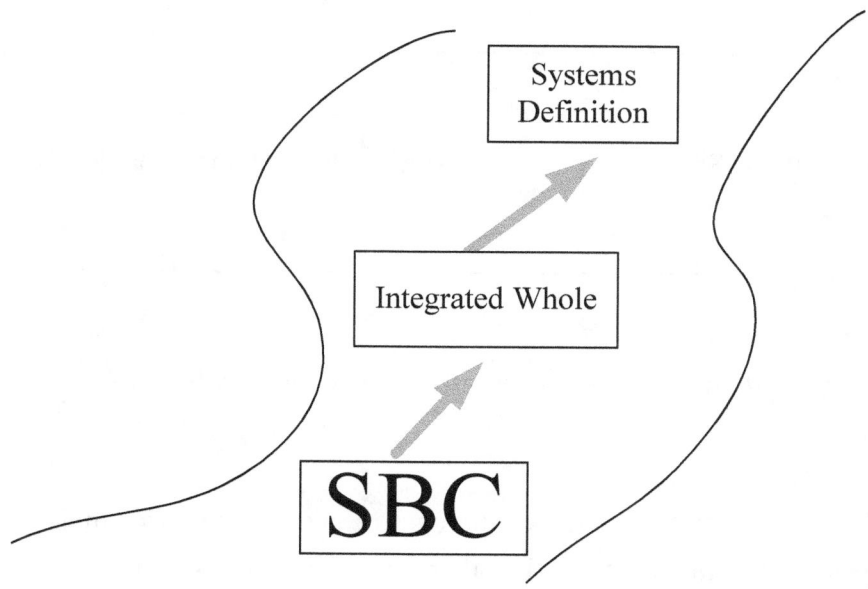

Figure 3-7   SBC to Achieve the Systems Definition

In the SBC approach, different systems structures may draw forth the same systems behavior as shown in Figure 3-8.

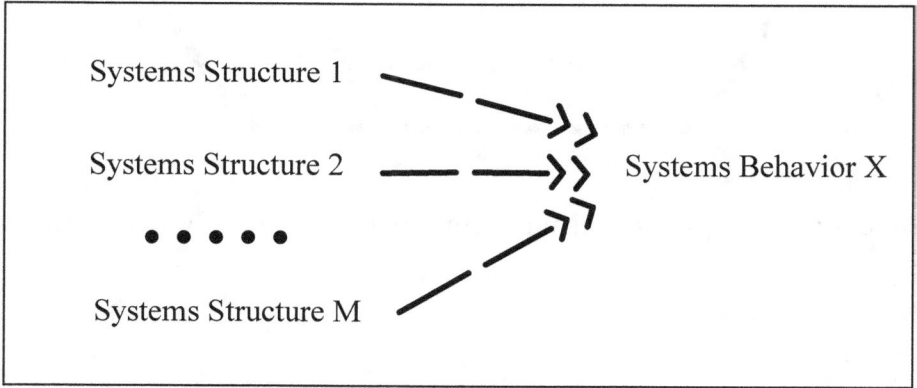

Figure 3-8   Different Systems Structures Draw Forth the Same Systems Behavior

Since there is only one systems structure exists in one systems definition, one systems behavior will always be attached to or built on one systems structure as shown in Figure 3-9.

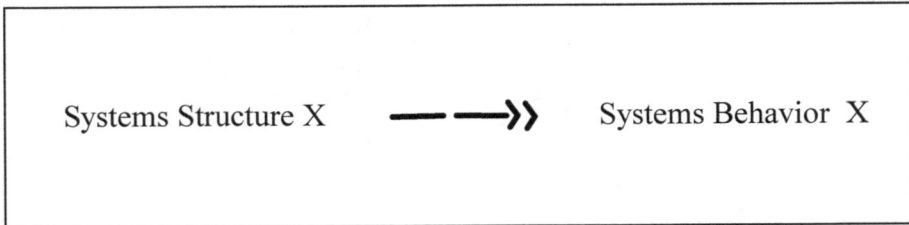

Figure 3-9   One Systems Behavior is Attached to One Systems Structure

We conclude that in the SBC approach, a systems behavior must be attached to or built on a systems structure. In other words, a systems behavior can not exist alone; it must be loaded on a systems structure just like a cargo is loaded on a ship as shown in Figure 3-10. There will be no systems behavior if there is no systems structure. A stand-alone systems behavior with no systems structure is not meaningful.

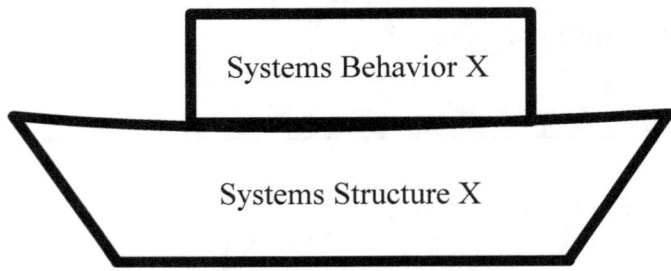

Figure 3-10   A Systems Behavior Must be Loaded on a Systems Structure

3-5 Systems Thinking 2.0

Since structure-behavior coalescence (SBC) provides an elegant way to integrate the systems structure and systems behavior, we shall include it in the definition of a system. Figure 3-11 shows how systems thinking 2.0 (architectural thinking) defines a system.

> A system,
> through the SBC architecture description language,
> truly is an integrated whole,
> embodied in its assembled components,
> their interactions with each other and the environment.

Figure 3-11   Systems Thinking 2.0 Defining a System

A system defined by systems thinking 2.0 (architectural thinking) has the following characteristics: 1) it emphasizes the system's structure-behavior coalescence; 2) it is a truly integrated whole; 3) it is embodied in its assembled components; 4) components are interacting (or handshaking) with each other and the environment; and 5) it uses structural decomposition [Chao12, Ghar11] rather than functional decomposition [Scho10].

Structure-behavior coalescence (SBC) provides an elegant way to integrate the systems structure and systems behavior of a system. Systems thinking 2.0 (architectural thinking) uses the SBC architecture description language (SBC-ADL) to formally describe the integration of systems structure and systems behavior of a system. SBC-ADL contains six fundamental diagrams: a) architecture hierarchy diagram, b) framework diagram, c) component operation diagram, d) component connection diagram, e) structure-behavior coalescence diagram and f) interaction flow diagram.

So far, we have introduced the systems thinking 2.0 (architectural thinking) which will be able to appropriately define a system. In the following chapters, we shall elaborate the details of SBC-ADL.

PART II: SBC ARCHITECTURE DESCRIPTION LANGUAGE

Chapter 4: Architecture Hierarchy Diagram

Systems thinking 2.0 (architectural thinking) uses an architecture hierarchy diagram (AHD) to define the multi-level decomposition and composition of a system. AHD is the first fundamental diagram to achieve structure-behavior coalescence.

4-1 Decomposition and Composition

The following is an example of systems decomposition and composition. The *Computer* system consists of *Monitor*, *Keyboard*, *Mouse* and *Case*, as shown in Figure 4-1. *Monitor*, *Keyboard*, *Mouse* and *Case* are subsystems that comprise the *Computer* system.

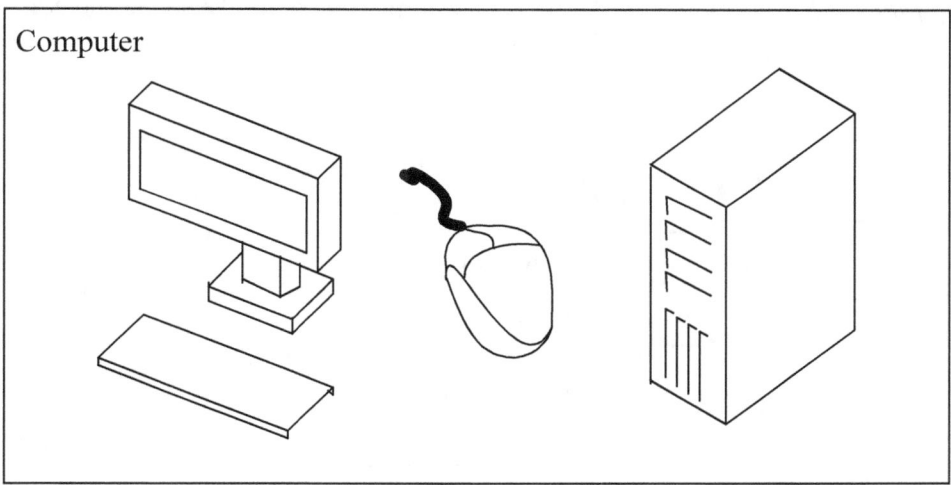

Figure 4-1   Decomposition and Composition of the *Computer* System

Another example indicates that the *Tree* system is composed of *Root* and *Stem*, as shown in Figure 4-2. In this example, we would say that *Root* and *Stem* are subsystems, respectively, while the *Tree* system consists of its subsystems.

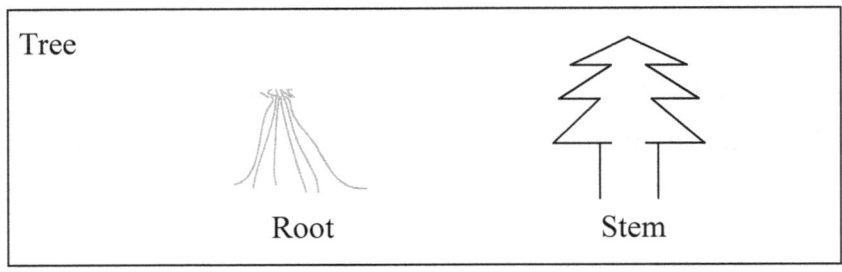

Figure 4-2   Decomposition and Composition of the *Tree* System

The last example demonstrates that the *SBC_Book* system is composed of *Chapter_1*, *Part_1* and *Part_2*, as shown in Figure 4-3. In this example, we would say that *Chapter_1*, *Part_1* and *Part_2* are subsystems, respectively while the *SBC_Book* system consists of its subsystems.

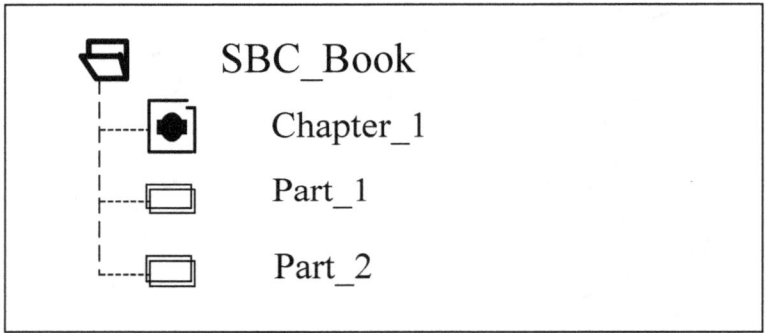

Figure 4-3   Decomposition and Composition of the *SBC_Book* System

Architecture hierarchy diagram (AHD) is used to define the decomposition and composition of a system. As an example, Figure 4-4 shows an AHD of the *Computer* system. We clearly observe that the *Computer* system is composed of *Monitor*, *Keyboard*, *Mouse* and *Case*.

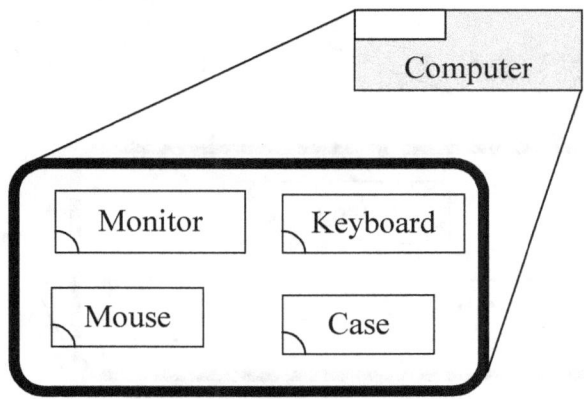

Figure 4-4   AHD of the *Computer* System

As a second example, Figure 4-5 shows an AHD of the *Tree* system. We clearly observe that the *Tree* system is composed of *Root* and *Stem*.

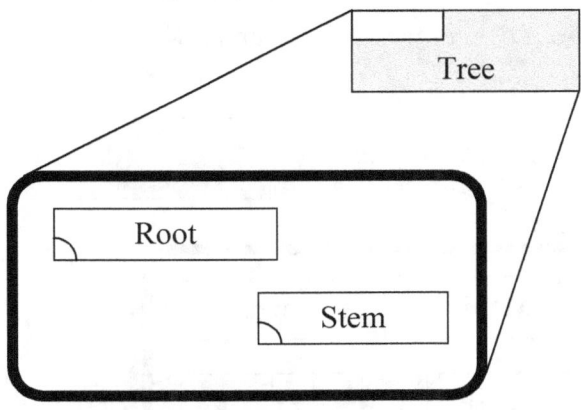

Figure 4-5   AHD of the *Tree* System

As a third example, Figure 4-6 shows an AHD of the *SBC_Book* system. We clearly observe that the *SBC_Book* is composed of *Chapter_1*, *Part_1* and *Part_2*.

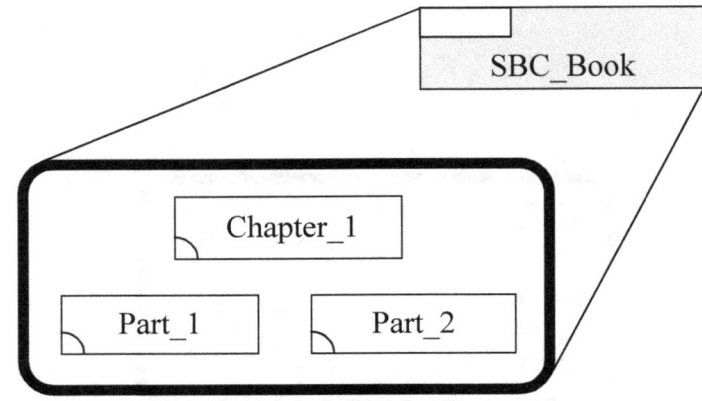

Figure 4-6   AHD of the *SBC_Book* system

4-2 Multi-Level Decomposition and Composition

The subsystem may also contain subsystems as we further decompose it. For example, *Case* is a subsystem of the *Computer*, and we can further decompose it into *Motherboard*, *Hard_Disk*, *Power_Supply* and *DVD_Disk*, as shown in Figure 4-7.

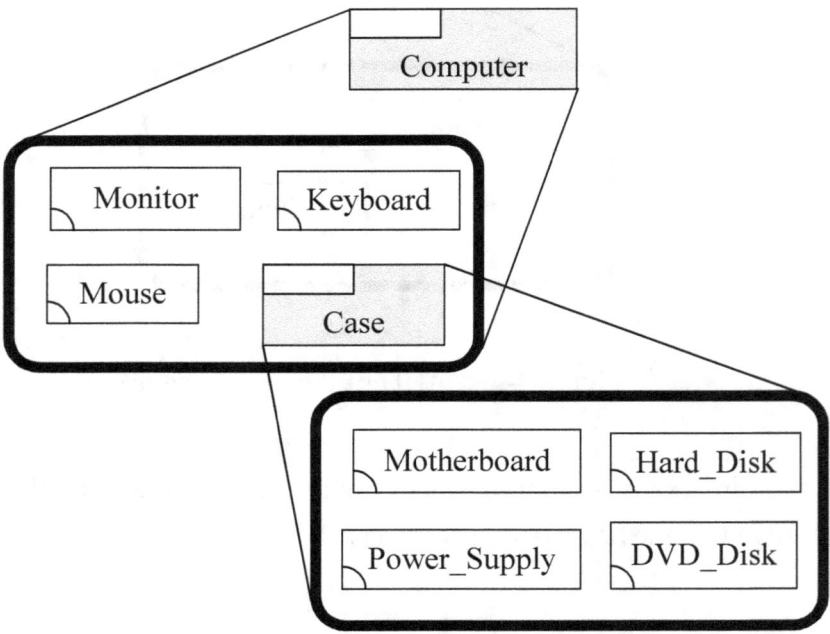

Figure 4-7   Multi-Level Decomposition/Composition of the *Computer* System

As a second example, *Stem* is a subsystem of the *Tree*, and we can further decompose it into *Trunk* and *Leaf*, as shown in Figure 4-8.

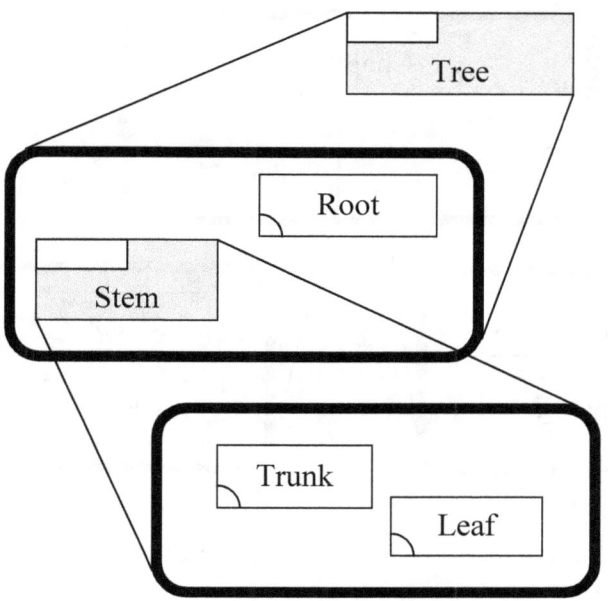

Figure 4-8   Multi-Level Decomposition/Composition of the *Tree* System

As a third example, *Part_1* is a subsystem of the *SBC_Book*, and we can further decompose it into *Chapter_2* and *Chapter_3*; *Part_2* is also a subsystem of the *SBC_Book*, and we can further decompose it into *Chapter_4* and *Chapter_5*, as shown in Figure 4-9.

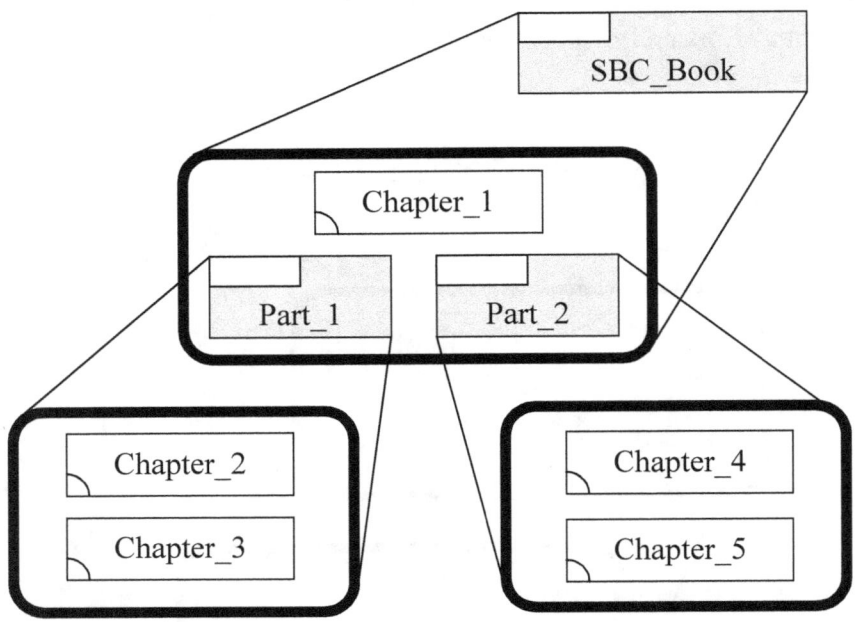

Figure 4-9   Multi-Level Decomposition/Composition of the *SBC_Book* System

Generally speaking, multi-level decomposition and composition of a system is applied often in defining a system. To make a complex system look simple, the mechanism of multi-level composition and decomposition should always be utilized.

4-3 Aggregated and Non-Aggregated Systems

Any subsystem (at any level) involved with multi-level decomposition and composition of a system is either aggregated or non-aggregated. The definition of aggregated and non-aggregated systems is shown in Figure 4-10.

Definition of Aggregated Systems:

A system (within an AHD) is aggregated if it is composed of any sub-system.

Definition of Non-aggregated Systems

A system (within an AHD) is non-aggregated if it is NOT composed of any sub-system.

Figure 4-10   Definition of Aggregated and Non-aggregated Systems

Non-aggregated systems are sometimes referred to as components, parts, entities, objects and building blocks [Chao09, Chao14].

In the multi-level systems decomposition and composition, any system is either aggregated or non-aggregated, but not both. For example, in Figure 4-4, *Case* is a non-aggregated system, not an aggregated system. As an interesting contrast, in Figure 4-7, *Case* is an aggregated system, not a non-aggregated system.

As a second example, in Figure 4-5, *Stem* is a non-aggregated system, not an aggregated system. As an interesting contrast, in Figure 4-8, *Stem* is an aggregated system, not a non-aggregated system.

As a third example, in Figure 4-6, *Part_1* and *Part_2* are non-aggregated systems, not aggregated systems. As an interesting contrast, in Figure 4-9, *Part_1* and *Part_2* are aggregated systems, not non-aggregated systems.

44

Chapter 5: Framework Diagram

Systems thinking 2.0 (architectural thinking) uses a framework diagram (FD) to define the multi-layer (also referred to as multi-tier) decomposition and composition of a system. FD is the second fundamental diagram to achieve structure-behavior coalescence.

5-1 Multi-Layer Decomposition and Composition

Decomposition and composition of a system can also be defined in a multi-layer manner. We draw a framework diagram (FD) for the multi-layer decomposition and composition of a system.

As an example, Figure 5-1 shows a FD of the *Computer* system. In the figure, *Technology_SubLayer_2* contains *Monitor*, *Keyboard* and *Mouse*; *Technology_SubLayer_1* contains *Motherboard*, *Hard_Disk*, *Power_Supply* and *DVD_Disk*.

Figure 5-1   FD of the *Computer* System

As a second example, Figure 5-2 shows a FD of the *Tree* system. In the figure, *Technology_SubLayer_2* contains *Root*; *Technology_SubLayer_1* contains *Trunk* and *Leaf*.

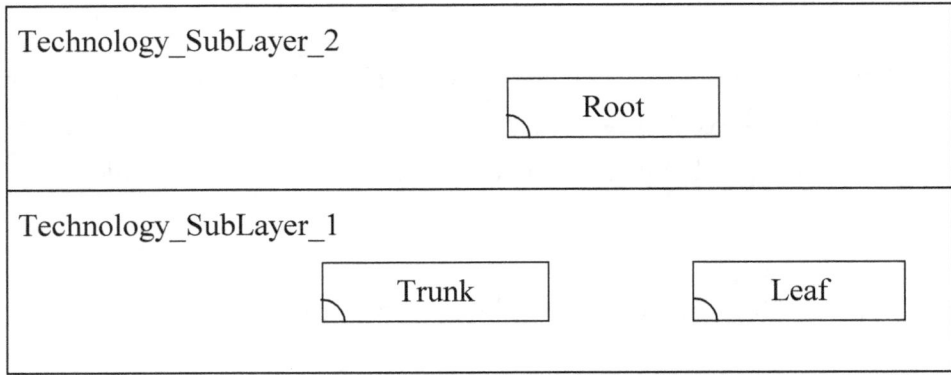

Figure 5-2   FD of the *Tree* System

As a third example, Figure 5-3 shows a FD of the *SBC_Book* system. In the figure, *Technology_SubLayer_2* contains *Chapter_1*; *Technology_SubLayer_1* contains *Chapter_2*, *Chapter_3*, *Chapter_4* and *Chapter_5*.

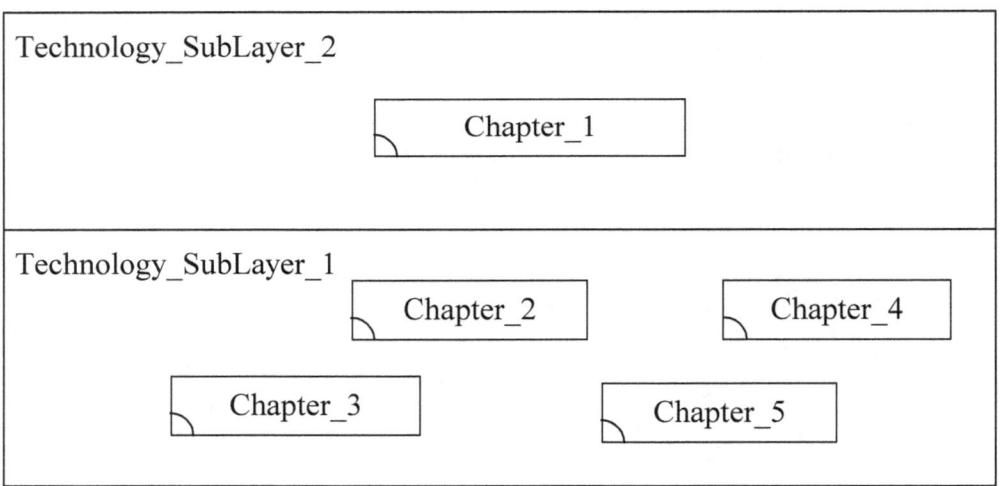

Figure 5-3   FD of the *SBC_Book* System

5-2 Only Non-Aggregated Systems Appearing in Framework Diagrams

Both aggregated and non-aggregated systems are displayed in the multi-level AHD decomposition and composition of a system. As an interesting contrast, only non-aggregated systems shall appear in the multi-layer FD decomposition and composition of a system.

For example, Figure 4-7 in the previous chapter shows an AHD of the *Computer* system in which both aggregated systems such as *Computer*, *Case* and non-aggregated systems such as *Monitor, Keyboard, Mouse, Motherboard, Hard_Disk, Power_Supply, DVD_Disk* are displayed. As an interesting contrast, Figure 5-1 in the previous section shows a FD of the *Computer* system in which only non-aggregated systems such as *Monitor, Keyboard, Mouse, Motherboard, Hard_Disk, Power_Supply* and *DVD_Disk* are displayed.

For a second example, Figure 4-8 in the previous chapter shows an AHD of the *Tree* system in which both aggregated systems such as *Tree*, *Stem* and non-aggregated systems such as *Root, Trunk, Leaf* are displayed. As an interesting contrast, Figure 5-2 in the previous section shows a FD of the *Tree* system in which only non-aggregated systems such as *Root, Trunk* and *Leaf* are displayed.

For a third example, Figure 4-9 in the previous chapter shows an AHD of the *SBC_Book* system in which both aggregated systems such as *SBC_Book, Part_1, Part_2* and non-aggregated systems such as *Chapter_1, Chapter_2, Chapter_3, Chapter_4, Chapter_5* are displayed. As an interesting contrast, Figure 5-3 in the previous section shows a FD of the *SBC_Book* system in which only non-aggregated systems such as *Chapter_1, Chapter_2, Chapter_3, Chapter_4* and *Chapter_5* are displayed.

48

Chapter 6: Component Operation Diagram

Systems thinking 2.0 (architectural thinking) uses a component operation diagram (COD) to define all components' operations of a system. COD is the third fundamental diagram to achieve structure-behavior coalescence.

6-1 Operations of Each Component

An operation provided by each component represents a procedure or method or function of the component. If other components request this component to perform an operation, then shall use it to accomplish the operation request.

Each component in a system must possess at least one operation. A component should not exist in a system if it does not possess any operation. Figure 6-1 shows that component *SalePurchase_GUI* has four operations: *SaleInputClick*, *SalePrintClick*, *PurchaseInputClick* and *PurchasePrintClick*.

Figure 6-1   Four Operations of *SalePurchase_GUI*

An operation formula is utilized to fully define an operation. An operation formula includes a) operation name, b) input parameters and c) output parameters as shown in Figure 6-2.

50

$$\text{Operation\_Name (In } a_1, a_2, ..., a_M \text{ ; Out } a_{M+1}, a_{M+2}, ..., a_{M+N} \text{)}$$

Figure 6-2   Operation Formula

Operation name is the name of this operation. In a system, every operation name should be unique. Duplicate operation names shall not be allowed in any system.

An operation may have several input and output parameters. The input and output parameters, gathered from all operations, represent the input data and output data views of a system [Date03, Elma10]. As shown in Figure 6-3, component *SalePrint_GUI* possesses the *ShowModal* operation which has no input/output parameter; component *SalePrint_GUI* also possesses the *SalePrintButtonClick* operation which has the *sDate* and *sNo* input parameters (with the arrow direction pointing to the component) and the *s_report* output parameter (with the arrow direction opposite to the component).

Figure 6-3   Input/Output Parameters of *SalePrintButtonClick*

Data formats of input and output parameters are defined by the data type specification. There are two sets of data types: primitive and composite [Date03, Elma10]. Figure 6-4 shows the primitive data type specification of the *sDate* and *sNo* input parameters occurring in the *SalePrintButtonClick(In sDate, sNo; Out s_report)* operation formula.

Parameter	Data Type	Instances
sDate	Text	20100517, 20100612
sNo	Text	001, 002

Figure 6-4   Primitive Data Type Specification

Figure 6-5 shows the composite data type specification of the *s_report* output parameter occurring in the *SalePrintButtonClick(In sDate, sNo; Out s_report)* operation formula.

Parameter	*s_report*
Data Type	TABLE of Sale Date : Text Sale No : Text Customer : Text ProductNo : Text Quantity : Integer UnitPrice : Real Total : Real End TABLE;
Instances	Sale Date : 20100517 Sale No : 001 Customer : Larry Fink <table><tr><td>ProductNo</td><td>Quantity</td><td>UnitPrice</td></tr><tr><td>A12345</td><td>400</td><td>100.00</td></tr><tr><td>A00001</td><td>300</td><td>200.00</td></tr></table>Total : 100,000.00

Figure 6-5   Composite Data Type Specification

6-2 Drawing the Component Operation Diagram

We use a component operation diagram (COD) to define all components' operations of a system. Figure 6-6 shows the *Multi-Tier Personal Data System*'s COD. In the figure, component *MTPDS_GUI* has two operations: *Calculate_AgeClick* and *Calculate_OverweightClick*; component *Age_Logic* has one operation: *Calculate_Age*; component *Overweight_Logic* has one operation: *Calculate_Overweight*; component *Personal_Database* has two operations: *Sql_DateOfBirth_Select* and *Sql_SexHeightWeight_Select*.

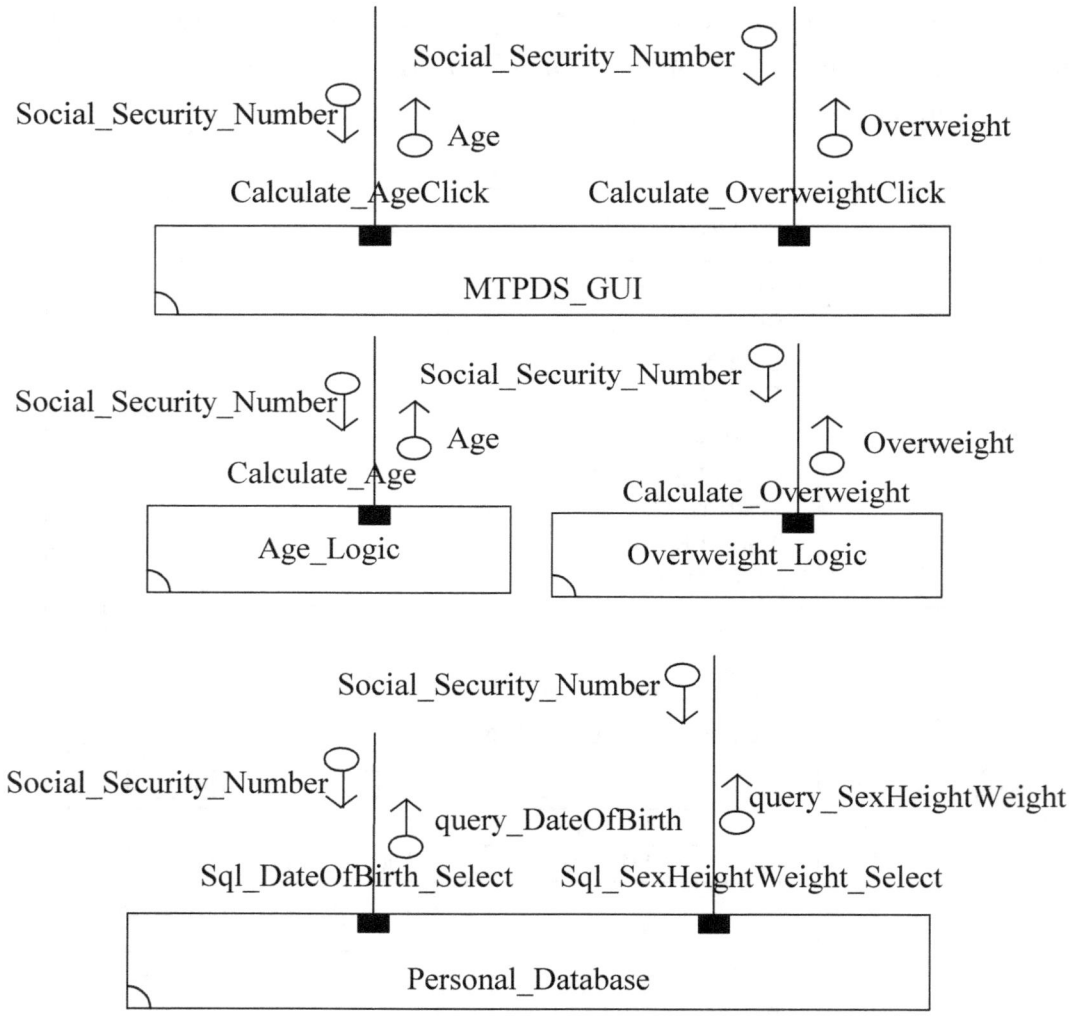

Figure 6-6   COD of the *Multi-Tier Personal Data System*

The operation formula of *Calculate_AgeClick* is *Calculate_AgeClick(In Social_Security_Number; Out Age)*. The operation formula of *Calculate_OverweightClick* is *Calculate_OverweightClick(In Social_Security_Number; Out Overweight)*. The operation formula of *Calculate_Age* is *Calculate_Age(In Social_Security_Number; Out Age)*. The operation formula of *Calculate_Overweight* is *Calculate_Overweight(In Social_Security_Number; Out Overweight)*. The operation formula of *Sql_DateOfBirth_Select* is *Sql_DateOfBirth_Select(In Social_Security_Number; Out query_DateOfBirth)*. The operation formula of *Sql_SexHeightWeight_Select* is *Sql_SexHeightWeight_Select(In Social_Security_Number; Out query_SexHeightWeight)*.

Figure 6-7 shows the primitive data type specification of the *Social_Security_Number* input parameter and the *Age*, *Overweight* output parameters.

Parameter	Data Type	Instances
Social_Security_Number	Text	424-87-3651, 512-24-3722
Age	Integer	28, 56
Overweight	Boolean	Yes, No

Figure 6-7   Primitive Data Type Specification

Figure 6-8 shows the composite data type specification of the *query_DateOfBirth* output parameter occurring in the *Sql_DateOfBirth_Select(In Social_Security_Number; Out query_DateOfBirth)* operation formula.

54

Parameter	*query_DateOfBirth*
Data Type	TABLE of Social_Security_Number : Text Age : Integer End TABLE ;
Instances	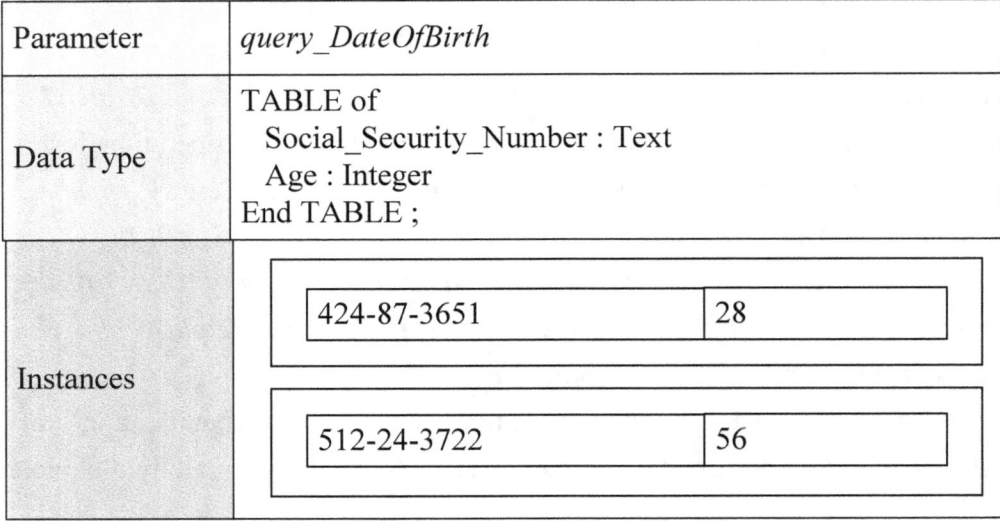

Figure 6-8   Composite Data Type Specification

Figure 6-9 shows the composite data type specification of the *query_SexHeightWeight* output parameter occurring in the *Sql_SexHeightWeight_Select(In Social_Security_Number; Out query_SexHeightWeight)* operation formula.

Parameter	*query_SexHeightWeight*			
Data Type	TABLE of Social_Security_Number : Text Sex : Text Height : Number Weight : Number End TABLE ;			
Instances	424-87-3651	Female	162	76
	512-24-3722	Male	180	80

Figure 6-9   Composite Data Type Specification

Chapter 7: Component Connection Diagram

Systems thinking 2.0 (architectural thinking) uses a component connection diagram (CCD) to define how all components and actors are connected within a system. CCD is the fourth fundamental diagram to achieve structure-behavior coalescence.

7-1 Essence of a Connection

A connection implies an operation request. When an operation is used by another subsystem then a connection appears. Accordingly, a connection is defined as the linkage that is constructed when an operation is used by another subsystem. Figure 7-1 shows that Subsystem_A uses the *Salary_Calculation* operation provided by the *Component_B* component.

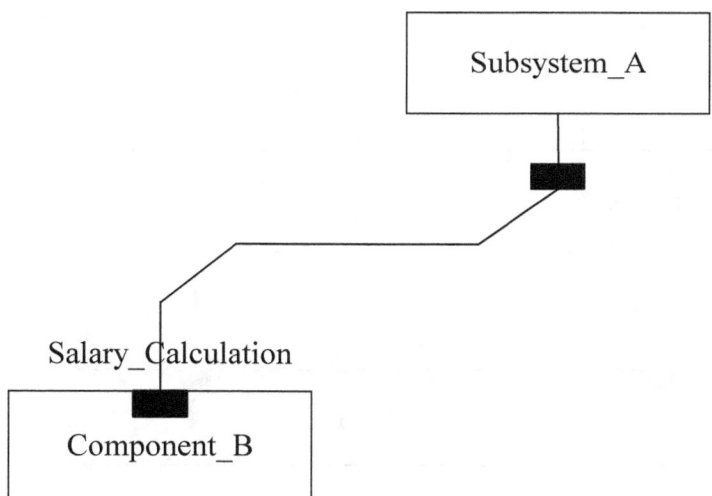

Figure 7-1   A Connection Appears When an Operation is Used

The above figure defines, sufficiently, the essence of a connection. However, we seldom use this kind of drawing. Instead, a simplified drawing of the above figure is often used as shown in Figure 7-2.

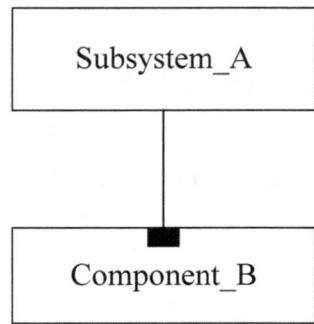

Figure 7-2   Simplified Drawing of a Connection

Since an operation is always provided by a component, there is no doubt that the *Component_B* operation provider is a component. On the contrary, the *Subsystem_A* operation user can be either a component (e.g., *Component_A*) or an actor (e.g., *Actor_A*) as shown in Figure 7-3. An actor belongs to the external environment of a system.

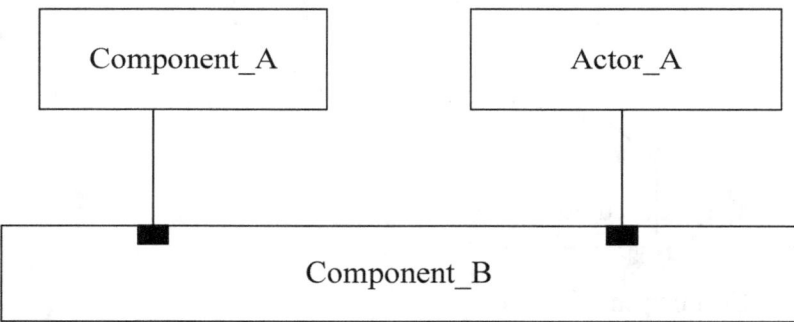

Figure 7-3   Operation User is Either a Component Or an Actor

Within a connection the subsystem (either a component or an actor) using the operation is always entitled the *Client* and the component which provides the operation is always entitled the *Server* as Figure 7-4 shows.

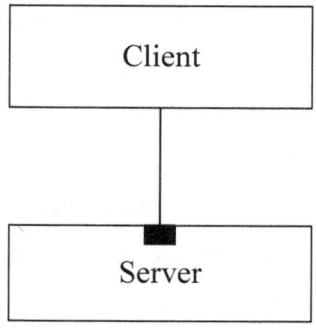

Figure 7-4   Roles of Client and Server Within a Connection

7-2 Drawing the Component Connection Diagram

A component connection diagram (CCD) is utilized to define how all components and actors (in the external environment) are connected within a system. Figure 7-5 exhibits the *Multi-Tier Personal Data System's COD*.

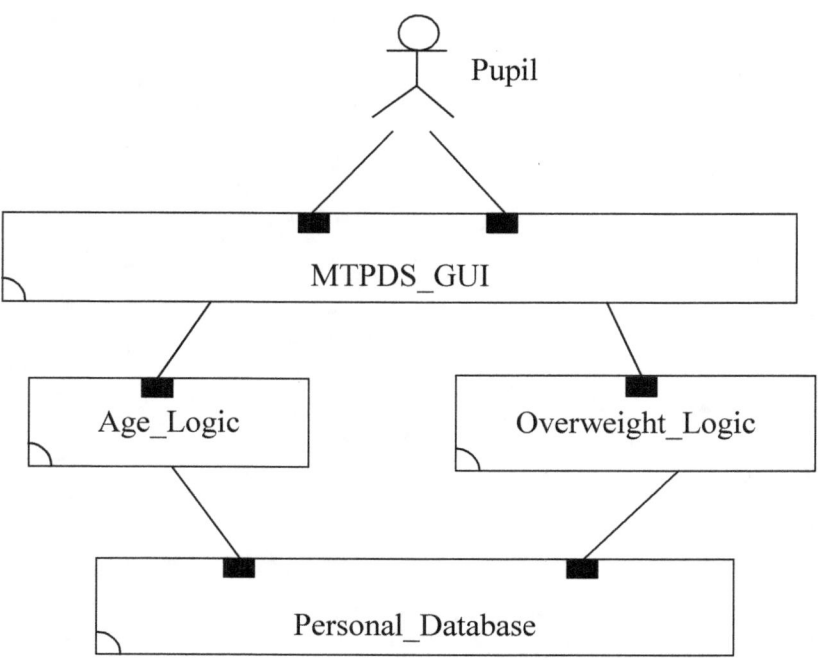

Figure 7-5   CCD of the *Multi-Tier Personal Data System*

In Figure 7-5, actor *Pupil* has two connections with the *MTPDS_GUI* component; component *MTPDS_GUI* has one connection with each of the *Age_Logic* and *Overweight_Logic* components; component *Age_Logic* has a connection with the *Personal_Database* component; component *Overweight_Logic* has a connection with the *Personal_Database* component.

After finishing the CCD, the formation pattern of the *Multi-Tier Personal Data System* will be constructed; thus the systems structure of the *Multi-Tier Personal Data System* becomes more transparent.

Chapter 8: Structure-Behavior Coalescence Diagram

Systems thinking 2.0 (architectural thinking) uses a structure-behavior coalescence diagram (SBCD) to define the systems structure and systems behavior coexisting in a system. SBCD is the fifth fundamental diagram to achieve structure-behavior coalescence.

8-1 Purpose of Structure-Behavior Coalescence Diagram

The major aim of SBC architecture description language is to achieve the integration of systems structure and systems behavior within a system. SBCD enables us to observe the systems structure and systems behavior coexisting in a system. This is the purpose of utilizing SBCD when defining a system.

Figure 8-1 exhibits a SBCD of the *Multi-Tier Personal Data System*. In this example, interactions (or handshakes) among the *Pupil* actor and the *MTPDS_GUI*, *Age_Logic*, *Overweight_Logic*, *Personal_Database* components shall draw forth the *AgeCalculation* and *OverweightCalculation* behaviors.

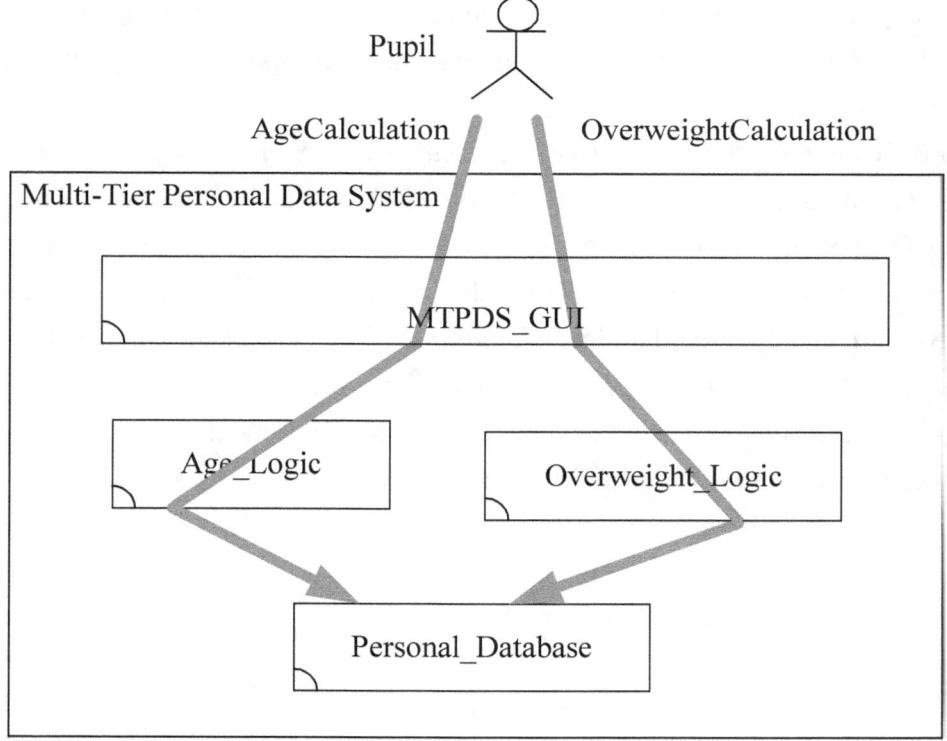

Figure 8-1   SBCD of the *Multi-Tier Personal Data System*

The overall behavior of a system is the aggregation of all its individual behaviors. All individual behaviors are mutually independent of each other. They tend to be executed concurrently [Hoar85, Miln89, Miln99]. For example, the overall behavior of the *Multi-Tier Personal Data System* includes the *AgeCalculation* and *OverweightCalculation* behaviors. In other words, the *AgeCalculation* and *OverweightCalculation* behaviors are combined to produce the overall behavior of the *Multi-Tier Personal Data System*.

The major purpose of using SBC architecture description language is to achieve the integration of systems structure and systems behavior within a system. In Figure 8-1, we are able to define the systems structure and systems behavior coexisting in a SBCD. That is, in the *Multi-Tier Personal Data System*'s SBCD, we not only see its systems structure but also see (at the same time) its systems behavior.

8-2 Drawing the Structure-Behavior Coalescence Diagram

Let us now explain the usage of SBCD by constructing a SBCD step by step. The goal of having a SBCD is enabling us to see both the systems structure and systems behavior, simultaneously. In order to achieve this goal, a SBCD is drawn by first defining all of the components, then defining the external environment's actors, and finally defining the interactions among these components and the external environment's actors.

For example, the *Multi-Tier Personal Data System* has two behaviors: *AgeCalculation* and *OverweightCalculation*. After constructing the *Multi-Tier Personal Data System* with all its components, the external environment's actors and the *AgeCalculation* behavior, we obtain the graphical representation as shown in Figure 8-2. In this Figure, the *AgeCalculation* behavior indicates that actor *Pupil* interacts with the *MTPDS_GUI* component first, then component *MTPDS_GUI* interacts with the *Age_Logic* component later, then component *Age_Logic* interacts with the *Personal_Database* component finally.

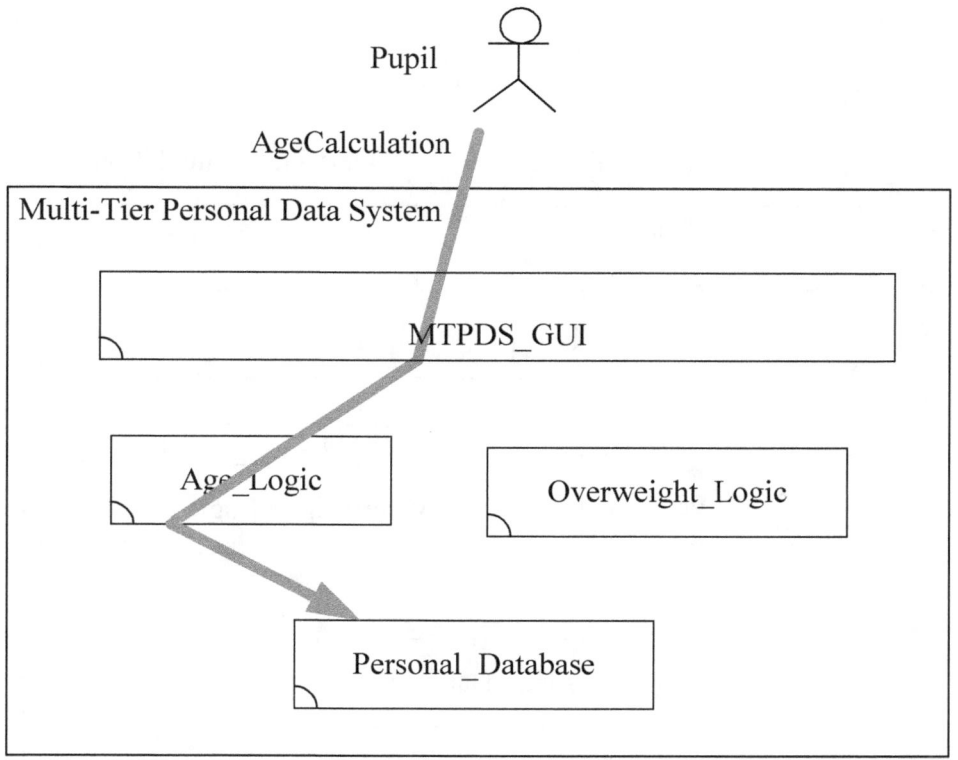

Figure 8-2   All Components, Actors, and the *AgeCalculation* Behavior

Adding the *OverweightCalculation* behavior to Figure 8-2, we then obtain the graphical representation shown in Figure 8-3. In this Figure, the *OverweightCalculation* behavior indicates that actor *Pupil* interacts with the *MTPDS_GUI* component first, then component *MTPDS_GUI* interacts with the *Overweight_Logic* component later, then component *Overweight_Logic* interacts with the *Personal_Database* component finally.

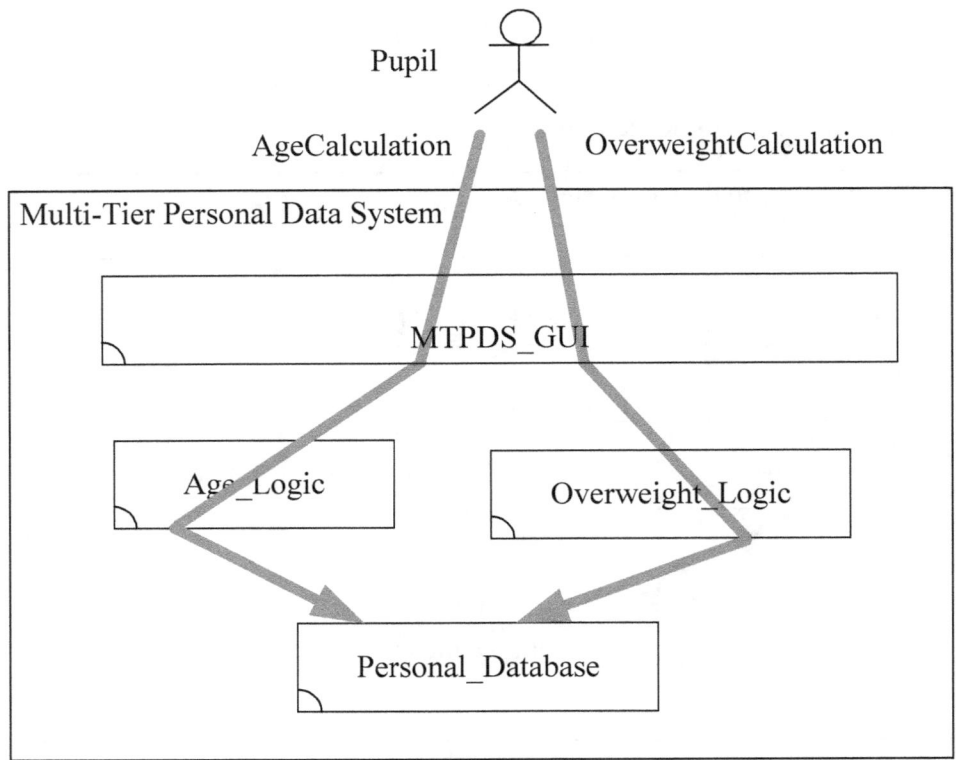

Figure 8-3   Adding the *OverweightCalculation* Behavior to Figure 8-2

After finishing Figure 8-3, we actually have accomplished all the works needed to draw an entire SBCD of the *Multi-Tier Personal Data System*. As a matter of fact, Figure 8-3 shows exactly the *Multi-Tier Personal Data System*'s SBCD.

Chapter 9: Interaction Flow Diagram

Systems thinking 2.0 (architectural thinking) uses an interaction flow diagram (IFD) to define each individual behavior of the overall behavior of a system. IFD is the sixth fundamental diagram to achieve structure-behavior coalescence.

9-1 Individual Behavior Represented by Interaction Flow Diagram

The overall behavior of a system consists of many individual behaviors. Each individual behavior represents an execution path. An IFD is utilized to define such an individual behavior.

Figure 9-1 demonstrates that the *Robot* system has two behaviors; thus, it has two IFDs.

System	IFD
Robot	Writing
	Walking

Figure 9-1   *Robot* System has Two IFDs

Figure 9-2 demonstrates that the *Multi-Tier Personal Data System* has two behaviors; thus, it has two IFDs.

System	IFD
Multi-Tier Personal Data System	AgeCalculation
	OverweightCalculation

Figure 9-2   *Multi-Tier Personal Data System* has Two IFDs

Figure 9-3 demonstrates that the *Sale and Purchase System* has four behaviors; thus, it has four IFDs.

System	IFD
Sale and Purchase System	SaleInput
	SalePrint
	PurchaseInput
	PurchasePrint

Figure 9-3   *Sale and Purchase System* has Four IFDs

Figure 9-4 demonstrates that the *Web Service Arithmetic System* has one behavior; thus it has one IFD.

System	IFD
Web Service Arithmetic System	Calculating (((P+Q)-R) *S)/T Value

Figure 9-4   *Web Service Arithmetic System* has One IFD

Figure 9-5 demonstrates that the *Web Service Arithmetic System* has one behavior; thus it has one IFD.

System	IFD
Web Service Extranet System	Purchase&Sale

Figure 9-5   *Web Service Extranet System* has One IFD

Figure 9-6 demonstrates that the *Convenience Store's Get 2nd 50% off Sales Promotion System* has two behaviors; thus it has two IFDs.

System	IFD
Convenience Store's Get 2nd 50% off Sales Promotion System	Advertising
	Get_2nd_50%_off

Figure 9-6   *Convenience Store's Get 2nd 50% off Sales Promotion System* has Two IFDs

Figure 9-7 demonstrates that the *Department Store's Car Sweepstakes Sales Promotion System* has three behaviors; thus it has three IFDs.

System	IFD
Department Store's Car Sweepstakes Sales Promotion System	Get_Sweepstake_Number
	Draw_Out_the_Winners_List
	Collect_the_Winning_Car

Figure 9-7   *Department Store's Car Sweepstakes Sales Promotion System* has Three IFDs

Figure 9-8 demonstrates that the *Female Mantis* has three behaviors; thus it has three IFDs.

System	IFD
Female Mantis	Preying,
	Walking
	Reproducing

Figure 9-8   *Female Mantis* has Three IFDs

Figure 9-9 demonstrates that the *Human Body* has three behaviors; thus it has three IFDs.

System	IFD
Human Body	Eating
	Respiring
	Combating

Figure 9-9   *Human Body* has Three IFDs

Figure 9-10 demonstrates that the *Disaster* has three behaviors; thus it has three IFDs.

System	IFD
	Flooding
Disaster	Catching_Fire
	Quaking

Figure 9-10   *Disaster* has Three IFDs

Figure 9-11 demonstrates that the *Bicycle* has three behaviors; thus it has three IFDs.

System	IFD
	Advancing
Bicycle	Turning_Left
	Braking

Figure 9-11   *Bicycle* has Three IFDs

Figure 9-12 demonstrates that the *Purchasing Management* has five behaviors; thus it has five IFDs.

System	IFD
Purchasing Management	Purchase_Requisition
	Quotation
	Purchase_Order
	Purchase
	Collect_Supplier_Data

Figure 9-12   *Purchasing Management* has Five IFDs

Figure 9-13 demonstrates that the *Restaurant* has one behavior; thus it has two IFDs.

System	IFD
Restaurant	Ordering_a_Meal
	Paying_the_Bill

Figure 9-13   *Restaurant* has One IFD

Figure 9-14 demonstrates that the *Car* has three behaviors; thus it has three IFDs.

System	IFD
Car	Setting_Gear_to_Forward
	Setting_Gear_to_Reverse
	Driving

Figure 9-14   *Car* has Three IFDs

9-2 Drawing the Interaction Flow Diagram

Let us now explain the usage of interaction flow diagram (IFD) by drawing an IFD step by step. Figure 9-15 demonstrates an IFD of the *AgeCalculation* behavior. The X-axis direction is from the left side to right side and the Y-axis direction is from the above to the below. Inside an IFD, there are four elements: a) external environment's actor, b) components, c) interactions and d) input/output parameters. Participants of the interaction, such as the external environment's actor and each component, are laid aside along the X-axis direction on the top of the diagram. The external environment's actor which initiates the sequential interactions is always placed on the most left side of the X-axis. Then, interactions among the external environment's actor and components successively in turn decorate along the Y-axis direction. The first interaction is placed on the top of the Y-axis position. The last interaction is placed on the bottom of the Y-axis position. Each interaction may carry several input and/or output parameters.

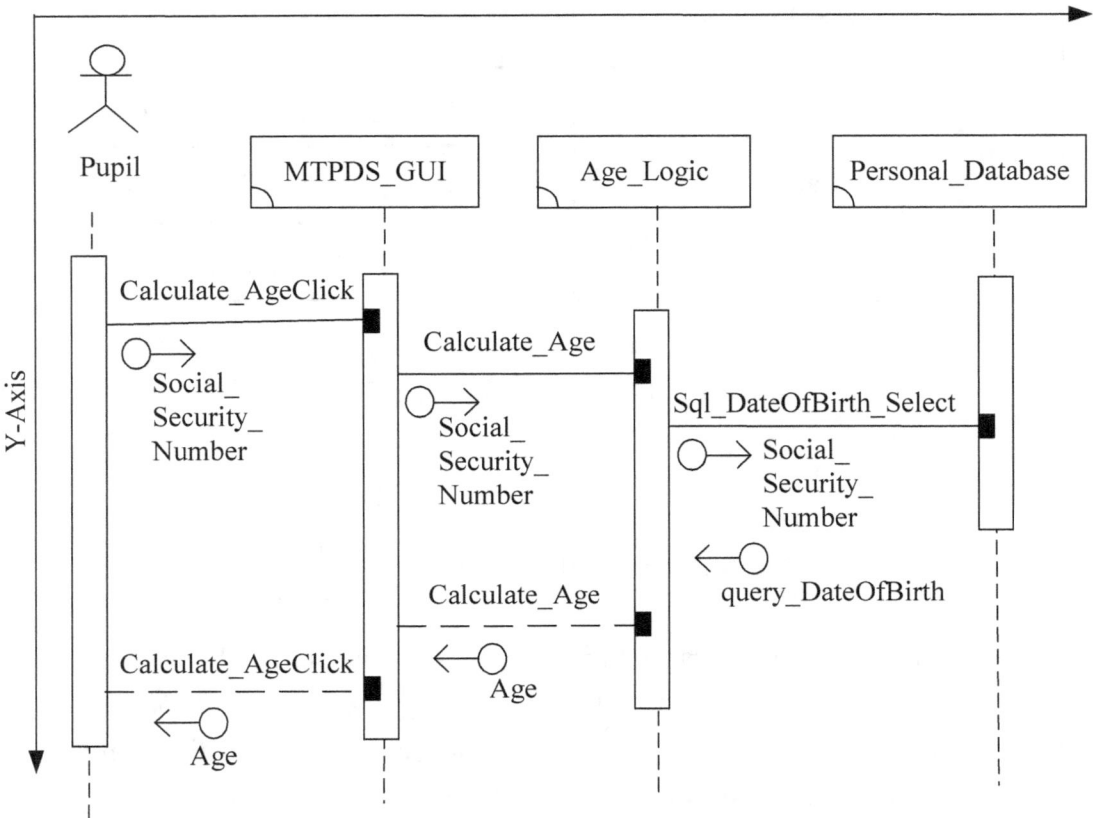

Figure 9-15   IFD of the *AgeCalculation* Behavior

In Figure 9-15, *Pupil* is an external environment's actor. *MTPDS_GUI*, *Age_Logic* and *Personal_Database* are components. *Calculate_AgeClick* is an operation provided by the *MTPDS_GUI* component, carrying the *Social_Security_Number* output parameter and *Age* input parameters. *Calculate_Age* is an operation provided by the *Age_Logic* component, carrying the *Social_Security_Number* input parameter and *Age* output parameter. *Sql_DateOfBirth_Select* is an operation provided by the component *Personal_Database*, carrying the *Social_Security_Number* input parameter and *query_DateOfBirth* output parameter.

The execution path of Figure 9-15 is as follows. First, actor *Pupil* interacts with the *MTPDS_GUI* component through the *Calculate_AgeClick* operation call interaction, carrying the *Social_Security_Number* input parameter. Next, component *MTPDS_GUI* interacts with the *Age_Logic* component through the *Calculate_Age* operation call interaction, carrying the *Social_Security_Number* input parameter.

Continuingly, component *Age_Logic* interacts with the *Personal_Database* component through the *Sql_DateOfBirth_Select* operation call interaction, carrying the *Social_Security_Number* input parameter and the *query_DateOfBirth* output parameter. Repeatedly, component *MTPDS_GUI* interacts with the *Age_Logic* component through the *Calculate_Age* operation return interaction, carrying the *Age* output parameter. Finally, actor *Pupil* interacts with the *MTPDS_GUI* component through the *Calculate_AgeClick* operation return interaction, carrying the *Age* output parameter.

For each interaction, the solid line stands for operation call while the dashed line stands for operation return. The operation call and operation return interactions, if using the same operation name, belong to the identical operation. Figure 9-16 exhibits two interactions (operation call interaction and operation return interaction) using the identical "*Request*" operation.

Figure 9-16   Two Interactions Have the Identical Operation

The execution path of Figure 9-16 is as follows. First, external environment's actor *Traveler* interacts with the *Agent* component through the *Request* operation call interaction, carrying the *Traveler_Name* input parameter. Next, component *Agent* interacts with the *Hotel* component through the *Reserve* operation call interaction, carrying the *Agent_Name* input parameter and *Hotel_Stauts* output parameter. Finally, external environment's actor *Traveler* interacts with the *Agent* component through the *Request* operation return interaction, carrying the *Agent_Stauts* output parameter.

An interaction flow diagram may contain a conditional expression. Figure 9-17 shows such an example which has the following execution path. First, external environment's actor *Employee* interacts with the *Computer* component through the *Open* operation call interaction, carrying the *Task_No* input parameter. Next, if the *var_1* < 4 & *var_2* > 7 condition is true then component *Computer* shall interact with the *Skype* component through the *Op_1* operation call interaction and component *Skype* shall interact with the *Earphone* component through the *Op_4* operation call interaction, carrying the *Skype_Earphone* output parameter; else if the *var_3* = 99 condition is true then component *Computer* shall interact with the *Skype* component through the *Op_2* operation call interaction and component *Skype* shall interact with the *Speaker* component through the *Op_5* operation call interaction, carrying the *Skype_Speaker* output parameter; else component *Computer* shall interact with the *Youtube* component through the *Op_3* operation call interaction and component *Youtube* shall interact with the *Speaker* component through the *Op_6* operation call interaction, carrying the *Youtube_Speaker* output parameter. Continuingly, if the *var_1* < 4 & *var_2* > 7 condition is true then component *Computer* shall interact with the *Skype* component through the *Op_1* operation return interaction, carrying the *Status_1* output parameter; else if the *var_3* = 99 condition is true then component *Computer* shall interact with the *Skype* component through the *Op_2* operation return interaction, carrying the *Status_2* output parameter; else component *Computer* shall interact with the *Youtube* component through the *Op_3* operation return interaction, carrying the *Status_3* output parameter. Finally, external environment's actor *Employee* interacts with the *Computer* component through the *Open* operation return interaction, carrying the *Status* output parameter.

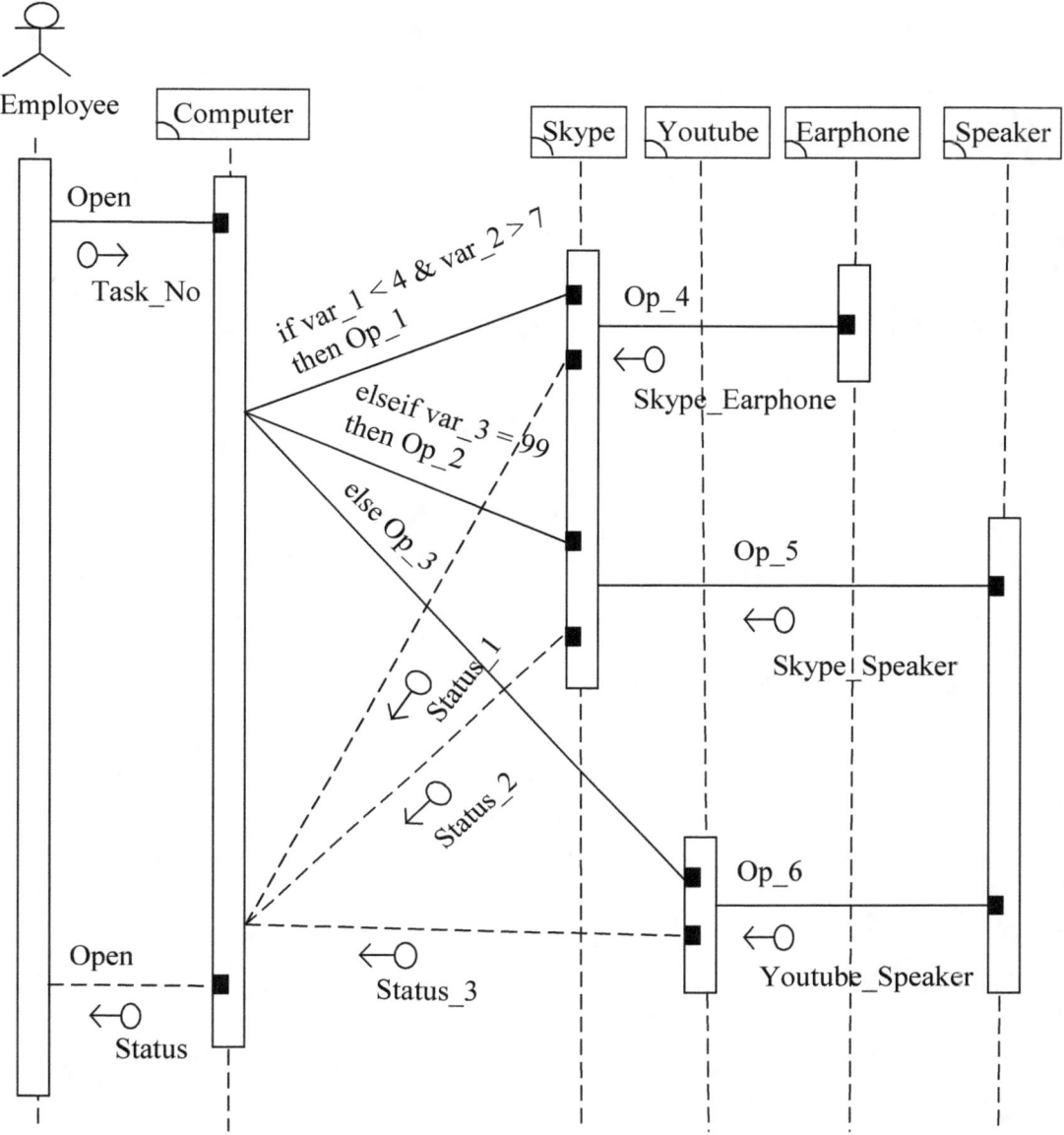

Figure 9-17   Conditional Interaction

Several Boolean conditions are shown in Figure 9-17. They are "*var_1 < 4 &*
var_2 > 7" and "*var_3 = 99*". Variables, such as *var_1*, *var_2* and *var_3*, appearing
in the Boolean condition can be local or global variables [Prat00, Seth96].

PART III: CASE STUDIES

Chapter 10: Systems Thinking 2.0 Defining the Robot System

This chapter demonstrates how to achieve systems thinking 2.0 (architectural thinking) defining the *robot* system, through the application of SBC architecture description language (SBC-ADL).

Usually, a person may use a remote controller to make the *robot* system write and walk. The overall behavior of the *robot* system is prominently represented by two individual behaviors: *Writing* and *Walking* as shown in Figure 10-1.

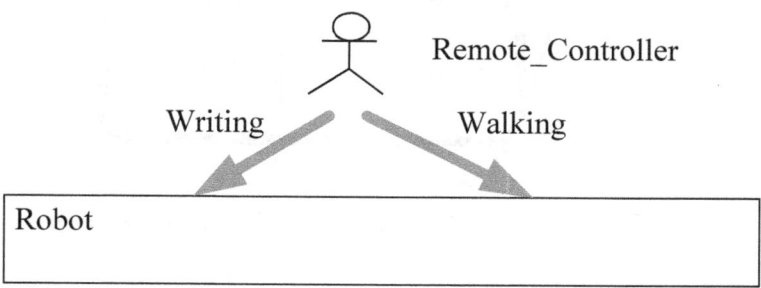

Figure 10-1   Two Behaviors of the *Robot* System

Using the SBC architecture description language, we shall go through: a) architecture hierarchy diagram, b) framework diagram, c) component operation diagram, d) component connection diagram, e) structure-behavior coalescence diagram and f) interaction flow diagram, to accomplish systems thinking 2.0 (architectural thinking) defining the *robot* system.

10-1 Architecture Hierarchy Diagram

We use an architecture hierarchy diagram (AHD) to define the multi-level composition and decomposition of the *robot* system. AHD is the first fundamental diagram to achieve structure-behavior coalescence. As shown in Figure 10-2, *Robot* is composed of *Head* and *Limb*; *Limb* is composed of *Hands* and *Feet*.

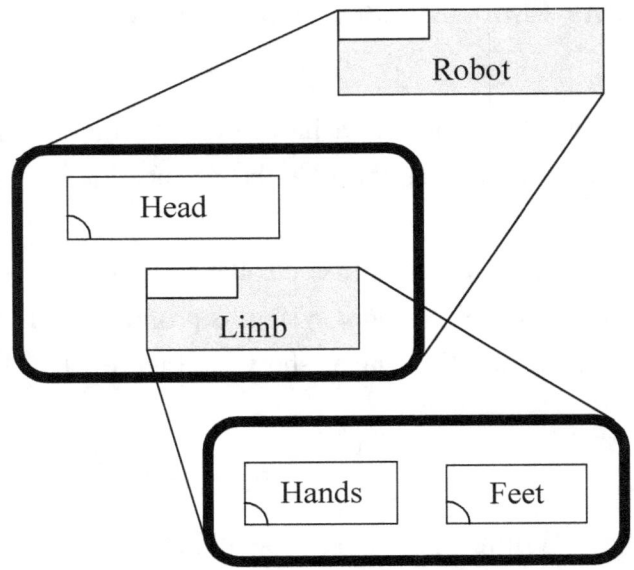

Figure 10-2   AHD of the *Robot* System

In Figure 10-2, *Robot* and *Limb* are aggregated systems while *Head*, *Hands* and *Feet* are non-aggregated systems.

10-2 Framework Diagram

We use a framework diagram (FD) to define the multi-layer composition and decomposition of the *robot* system as shown in Figure 10-3. FD is the second fundamental diagram to achieve structure-behavior coalescence. In the figure, *Technology_SubLayer_2* contains the *Head* component; *Technology_SubLayer_1* contains the *Hands* and *Feet* components.

Figure 10-3   FD of the *Robot* Sytsem

10-3 Component Operation Diagram

We use a component operation diagram (COD) to define the operations of all components of the *robot* system as shown in Figure 10-4. COD is the third fundamental diagram to achieve structure-behavior coalescence. In the figure, component *Head* has two operations: *Receive_Write_Signal* and *Receive_Walk_Signal*; component *Hands* has one operation: *Move_Hand*; component *Feet* has one operation: *Move_Foot*.

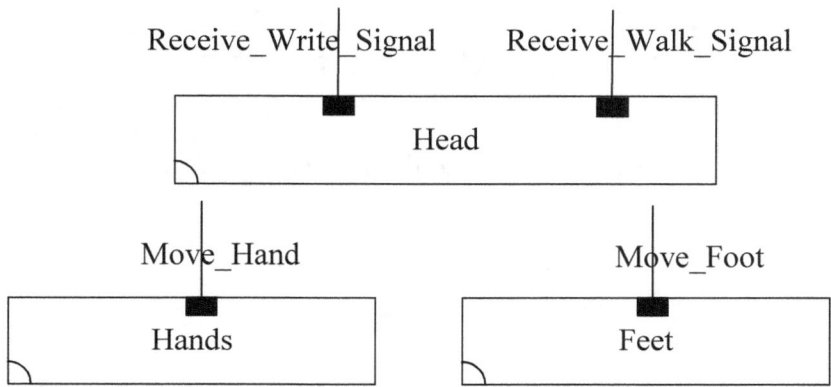

Figure 10-4　COD of the *Robot* System

10-4 Component Connection Diagram

We use a component connection diagram (CCD) to define the connections among components and actors of the *robot* system as shown in Figure 10-5. CCD is the fourth fundamental diagram to achieve structure-behavior coalescence.

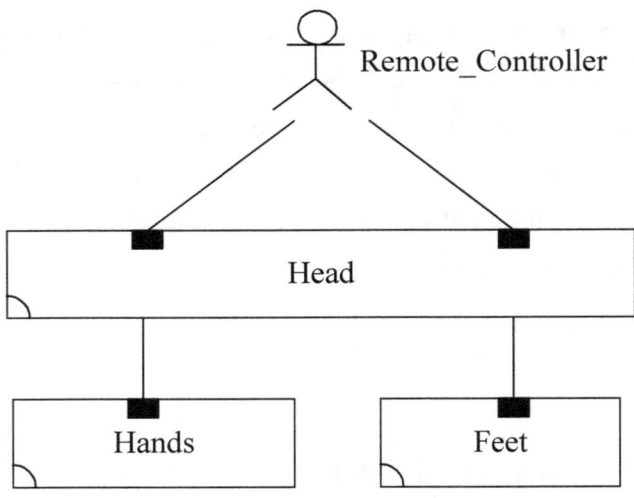

Figure 10-5   CCD of the *Robot* System

In Figure 10-5, actor *Remote_Controller* has a connection with the *Head* component; component *Head* has a connection with each of the *Hands* and *Feet* components.

10-5 Structure-Behavior Coalescence Diagram

We use a structure-behavior coalescence diagram (SBCD) to define the systems structure and systems behavior coexisting in the *robot* system as shown in Figure 10-6. SBCD is the fifth fundamental diagram to achieve structure-behavior coalescence. In the figure, interactions among the *Remote_Controller* actor and the *Head*, *Hands*, *Feet* components shall draw forth the *Writing* and *Walking* behaviors.

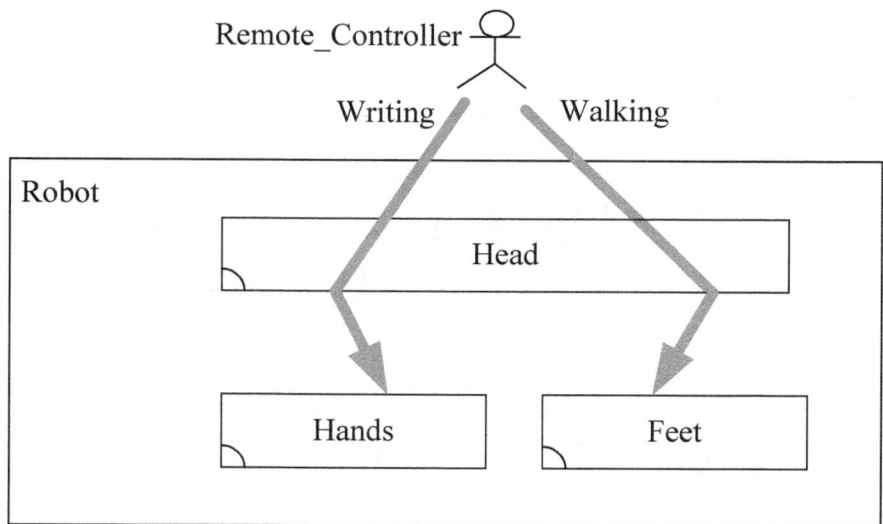

Figure 10-6   SBCD of the *Robot* System

The overall behavior of a system is the aggregation of all its individual behaviors. For example, the overall behavior of the *robot* system includes the *Writing* and *Walking* behaviors.

Be noticed that the *Writing* and *Walking* behaviors are mutually independent of each other. They tend to be executed concurrently [Hoar85, Miln89, Miln99].

The major purpose of SBC architecture description language is to define an integrated whole of a system. In Figure 10-6, we not only see its structure, but also see at the same time its behavior in the *robot* system's SBCD.

10-6 Interaction Flow Diagram

The overall behavior of the *robot* system includes two individual behaviors: *Writing* and *Walking*. Each individual behavior is represented by an execution path. We use an interaction flow diagram (IFD) to define each one of these execution paths. IFD is the sixth fundamental diagram to achieve structure-behavior coalescence.

Figure 10-7 shows an IFD of the *Writing* behavior. First, actor *Remote_Controller* interacts with the *Head* component through the *Receive_Write_Signal* operation call interaction. Finally, component *Head* interacts with the *Hands* component through the *Move_Hand* operation call interaction.

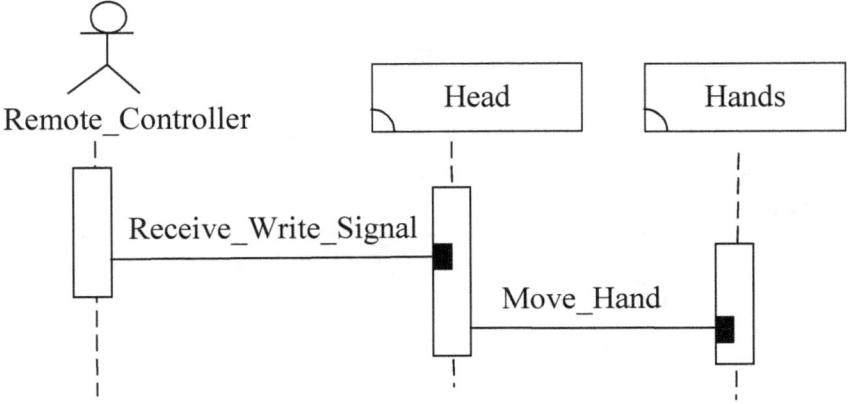

Figure 10-7   IFD of the *Writing* Behavior

Figure 10-8 shows an IFD of the *Walking* behavior. First, actor *Remote_Controller* interacts with the *Head* component through the *Receive_Walk_Signal* operation call interaction. Finally, component *Head* interacts with the *Feet* component through the *Move_Foot* operation call interaction.

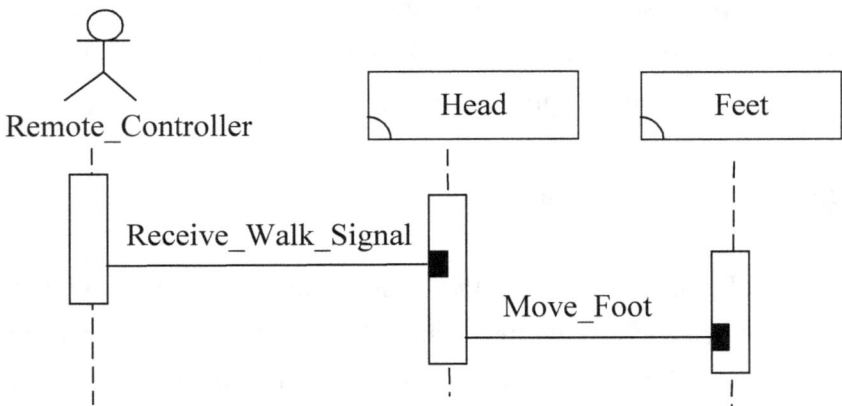

Figure 10-8   IFD of the *Walking* Behavior

Chapter 11: Systems Thinking 2.0 Defining the Multi-Tier Personal Data System

This chapter demonstrates how to achieve systems thinking 2.0 (architectural thinking) defining the *Multi-Tier Personal Data System*, through the application of SBC architecture description language (SBC-ADL).

After the systems development is finished, the *Multi-Tier Personal Data System* shall appear on a multi-tier platform [Wall04] as shown in Figure 11-1.

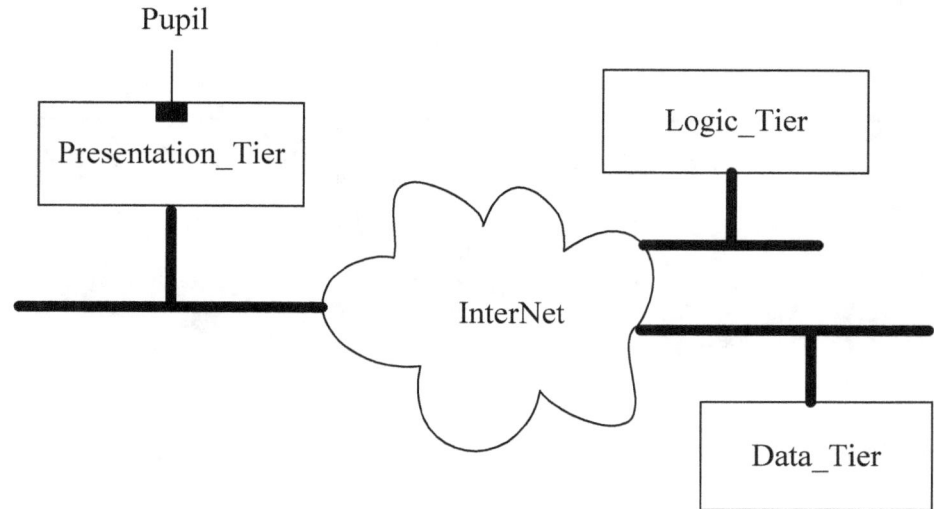

Figure 11-1   *Multi-Tier Personal Data System* on a Multi-Tier Platform

In the *Data_Tier*, there is a *Personal_Database* database [Date03, Elma10] which contains a *Personal_Data* table as shown in Figure 11-2.

Personal_Database : Database

Personal_Data : Table

Social_Secuirty_Number	Name	Date_of_Birth	Sex	Height (cm)	Weight (Kg)
318-49-2465	Mary R. Williams	June 17, 1976	Female	165	51
424-87-3651	Lee H. Wulf	July 24, 1982	Female	162	76
512-24-3722	John K. Bryant	May 12, 1954	Male	180	80

Record: ◄◄ ◄ | 4   ► ►◄ ►* of 4

Figure 11-2   *Personal_Database* Contains *Personal_Data*

The functionality of the *Multi-Tier Personal Data System* is to provide a graphical user interface (GUI) [Gali07] for the *Pupil* actor to trigger two behaviors. The first behavior is *AgeCalculation* and the second behavior is *OverweightCalculation*, as shown in Figure 11-3.

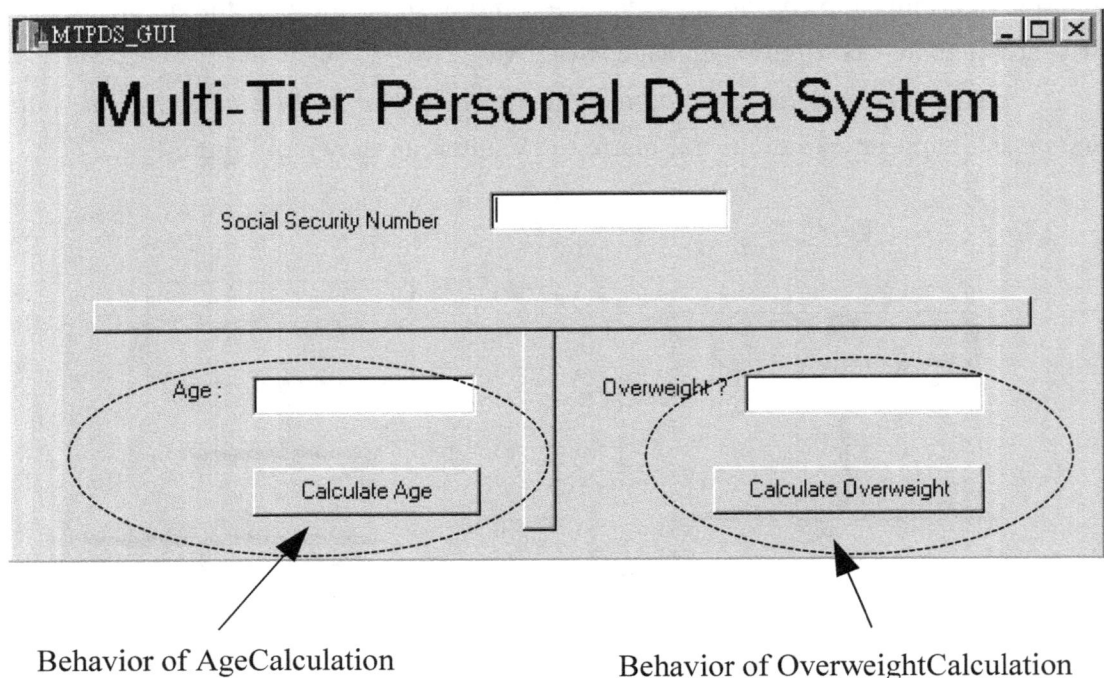

Figure 11-3   Two Behaviors

In the *AgeCalculation* behavior, actor *Pupil* inputs an integer *Social_Security_Number* value then presses down the *Calculate_Age* button. After that, the *Multi-Tier Personal Data System* retrieves the *Date_of_Birth* value from the database in line with the corresponding *Social_Security_Number* value. From the *Date_of_Birth* value, the *Multi-Tier Personal Data System* calculates the *Age* value and displays it on the screen. Figure 19-4 shows the *Social_Security_Number* value is 512-24-3722 and the retrieved *Date_of_Birth* value is May 12, 1954 and the calculated *Age* value, which is 59, is then displayed on the screen.

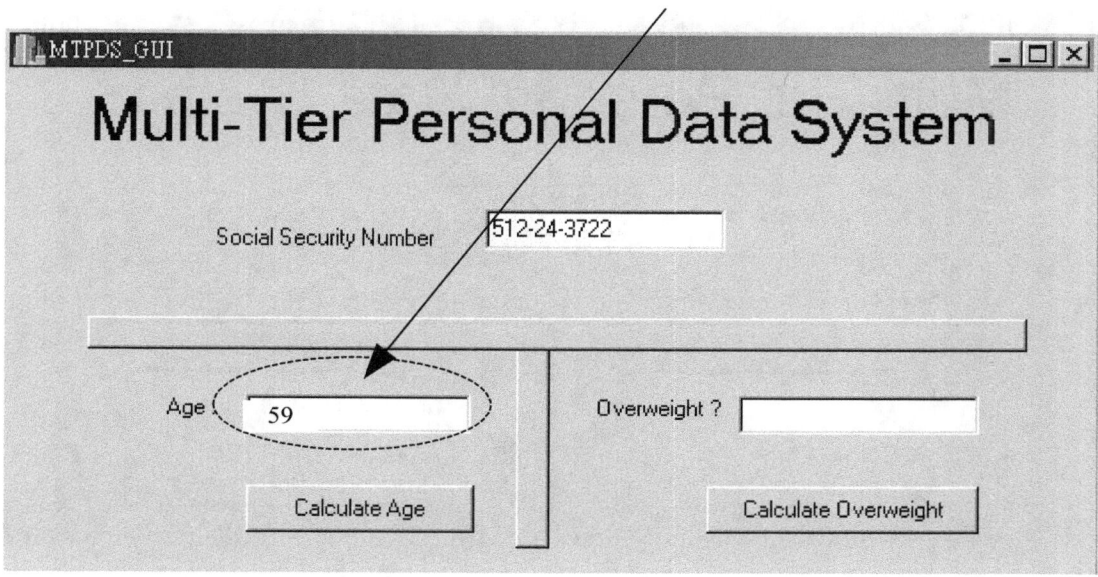

Figure 11-4   Behavior of *AgeCalculation*

In the *OverweightCalculation* behavior, actor *Pupil* inputs an integer *Social_Security_Number* value then presses down the *Calculate_Overweight* button. After that, the *Multi-Tier Personal Data System* retrieves the *Weight*, *Height* and *Sex* values from the database in line with the corresponding *Social_Security_Number* value. From the *Weight*, *Height* and *Sex* values, the *Multi-Tier Personal Data System* calculates the true-or-false *Overweight* value and displays it on the screen. Figure 19-5 shows the *Social_Security_Number* value is 318-49-2465 and the retrieved Sex, Height and *Weight* values are Female, 165 and 51, respectively, the calculated *Overweight* value, which is *No*, is then displayed on the screen.

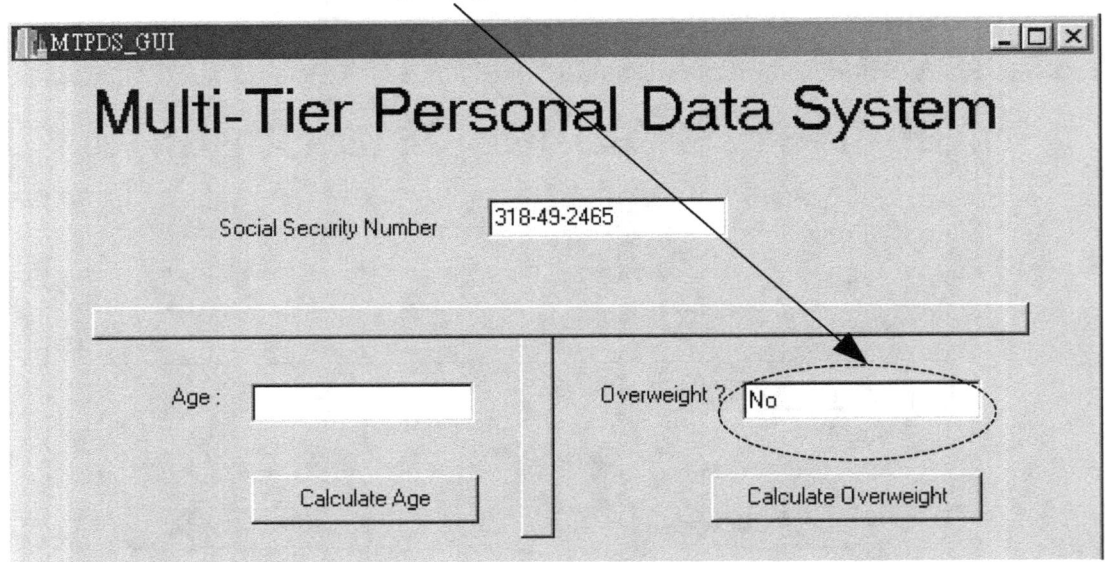

Calculated Overweight when Sex, Height (cm),
and Weight (kg) are Female, 165, and 51

Figure 11-5   Behavior of *OverweightCalculation*

Using the SBC architecture description language, we shall go through: a) architecture hierarchy diagram, b) framework diagram, c) component operation diagram, d) component connection diagram, e) structure-behavior coalescence diagram and f) interaction flow diagram, to accomplish systems thinking 2.0 (architectural thinking) defining the *Multi-Tier Personal Data System*.

11-1 Architecture Hierarchy Diagram

We use an architecture hierarchy diagram (AHD) to define the multi-level composition and decomposition of the *Multi-Tier Personal Data System*. AHD is the first fundamental diagram to achieve structure-behavior coalescence. As shown in Figure 11-6, *Multi-Tier Personal Data System* is composed of *MTPDS_GUI* and *M_Subsystem_2*; *M_Subsystem_2* is composed of *Age_Logic*, *Overweight_Logic* and

M_Subsystem_1; *M_Subsystem_1* is composed of *Personal_Database*.

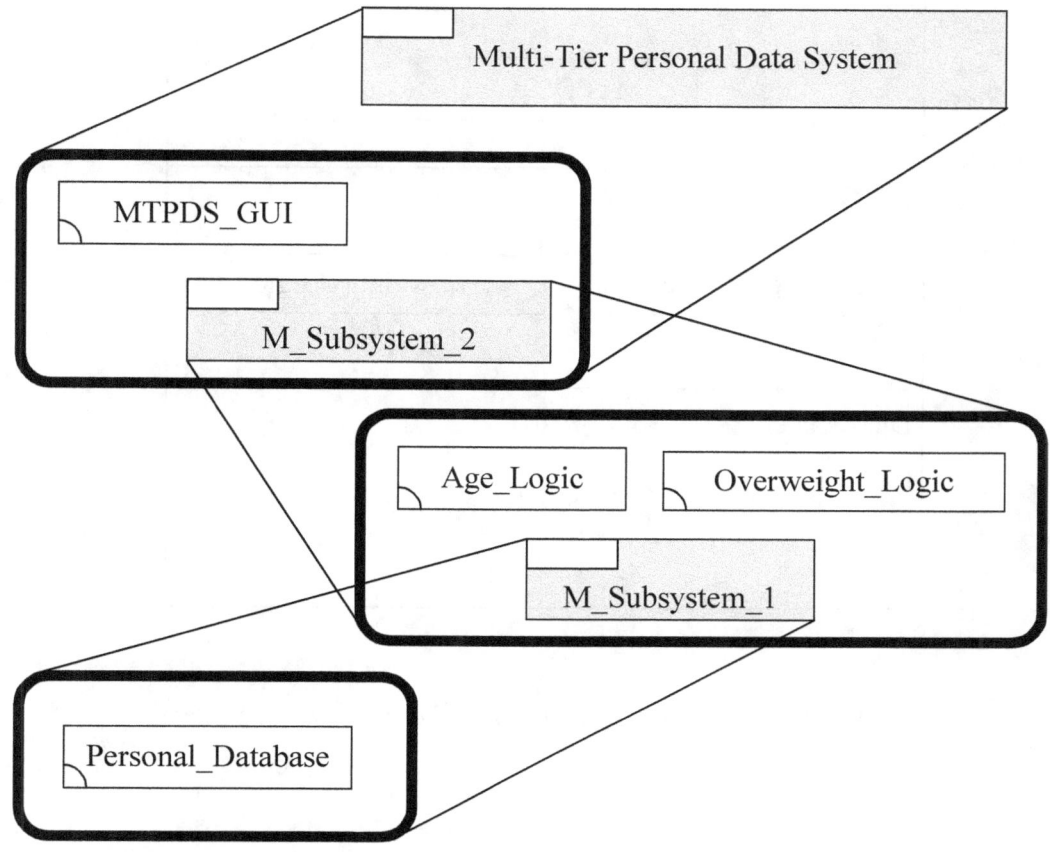

Figure 11-6   AHD of the Multi-Tier Personal Data System

In Figure 11-6, *Multi-Tier Personal Data System*, *M_Subsystem_2* and *M_Subsystem_1* are aggregated systems while *MTPDS_GUI*, *Age_Logic*, *Overweight_Logic* and *Personal_Database* are non-aggregated systems.

11-2 Framework Diagram

We use a framework diagram (FD) to define the multi-layer composition and decomposition of the *Multi-Tier Personal Data System* as shown in Figure 11-7. FD is the second fundamental diagram to achieve structure-behavior coalescence. In the figure, *Application_Layer* contains the *MTPDS_GUI* component; *Logic_Layer* contains the *Age_Logic* and *Overweight_Logic* components; *Data_Layer* contains the *Personal_Database* component.

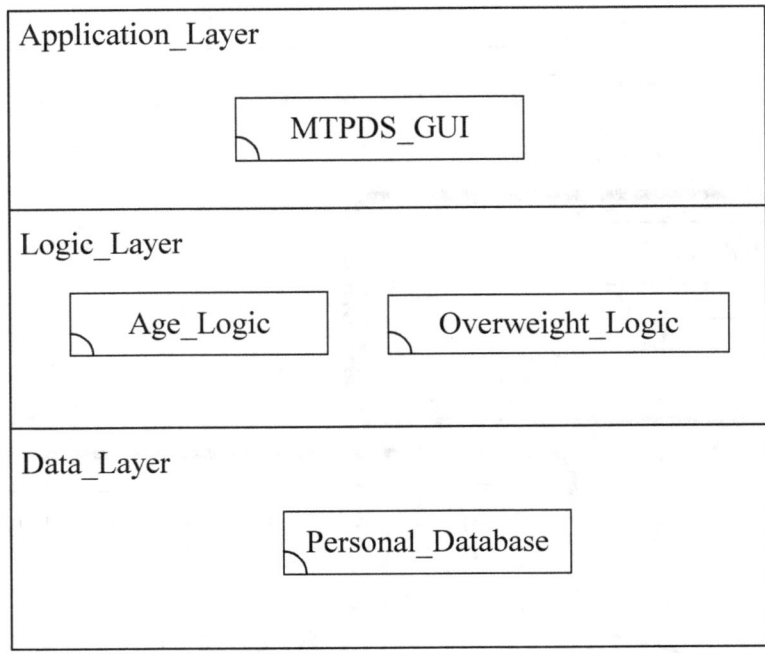

Figure 11-7   FD of the *Multi-Tier Personal Data System*

11-3 Component Operation Diagram

We use a component operation diagram (COD) to define the operations of all components of the *Multi-Tier Personal Data System* as shown in Figure 11-8. COD is the third fundamental diagram to achieve structure-behavior coalescence. In the figure, component *MTPDS_GUI* has two operations: *Calculate_AgeClick* and *Calculate_OverweightClick*; component *Age_Logic* has one operation: *Calculate_Age*; component *Overweight_Logic* has one operation: *Calculate_Overweight*; component *Personal_Database* has two operations: *Sql_DateOfBirth_Select* and *Sql_SexHeightWeight_Select*.

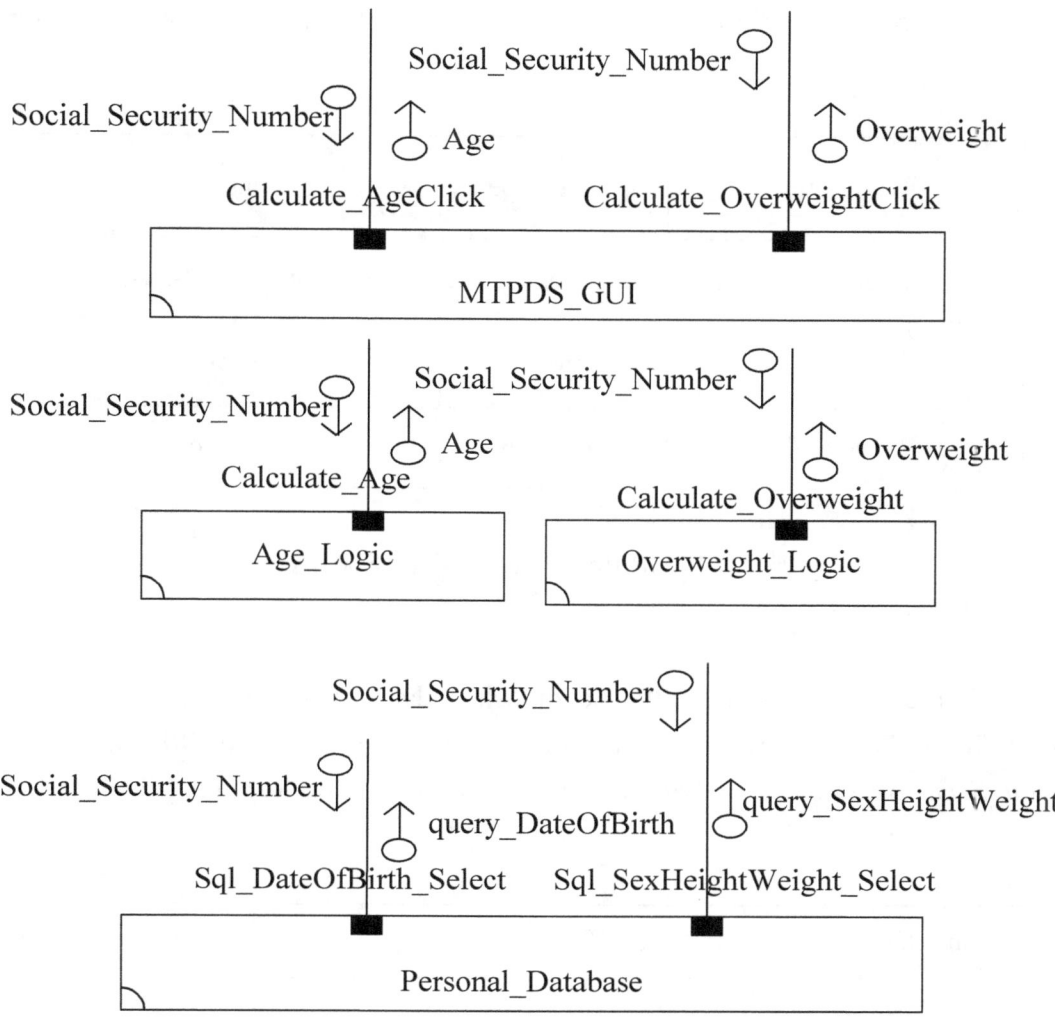

Figure 11-8   COD of the *Multi-Tier Personal Data System*

The operation formula of *Calculate_AgeClick* is *Calculate_AgeClick(In Social_Security_Number; Out Age)*. The operation formula of *Calculate_OverweightClick* is *Calculate_OverweightClick(In Social_Security_Number; Out Overweight)*. The operation formula of *Calculate_Age* is *Calculate_Age(In Social_Security_Number; Out Age)*. The operation formula of *Calculate_Overweight* is *Calculate_Overweight(In Social_Security_Number; Out Overweight)*. The operation formula of *Sql_DateOfBirth_Select* is *Sql_DateOfBirth_Select(In Social_Security_Number; Out query_DateOfBirth)*. The operation formula of *Sql_SexHeightWeight_Select* is *Sql_SexHeightWeight_Select(In Social_Security_Number; Out query_SexHeightWeight)*.

Figure 11-9 shows the primitive data type specification of the *Social_Security_Number* input parameter and the *Age*, *Overweight* output parameters.

Parameter	Data Type	Instances
Social_Security_Number	Text	424-87-3651, 512-24-3722
Age	Integer	28, 56
Overweight	Boolean	Yes, No

Figure 11-9   Primitive Data Type Specification

Figure 11-10 shows the composite data type specification of the *query_DateOfBirth* output parameter occurring in the *Sql_DateOfBirth_Select(In Social_Security_Number; Out query_DateOfBirth)* operation formula.

Parameter	*query_DateOfBirth*

Data Type	TABLE of
	   Social_Security_Number : Text
	   Age : Integer
	End TABLE;

Instances	424-87-3651 28 512-24-3722 56

Figure 11-10   Composite Data Type Specification

Figure 11-11 shows the composite data type specification of the *query_SexHeightWeight* output parameter occurring in the *Sql_SexHeightWeight_Select(In Social_Security_Number; Out query_SexHeightWeight)* operation formula.

Parameter	*query_SexHeightWeight*			

Data Type	TABLE of
	    Social_Security_Number : Text
	    Sex : Text
	    Height : Number
	    Weight : Number
	End TABLE;			

Instances	424-87-3651	Female	162	76
	512-24-3722	Male	180	80

Figure 11-11   Composite Data Type Specification

11-4 Component Connection Diagram

We use a component connection diagram (CCD) to define the connections among components and actors of the *Multi-Tier Personal Data System* as shown in Figure 11-12. CCD is the fourth fundamental diagram to achieve structure-behavior coalescence.

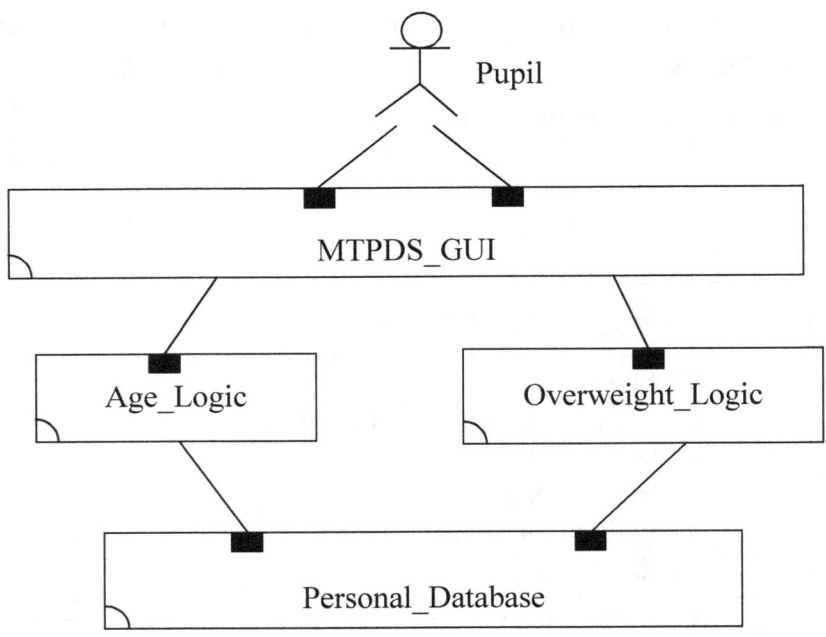

Figure 11-12   CCD of the *Multi-Tier Personal Data System*

In Figure 11-12, actor *Pupil* has two connections with the *MTPDS_GUI* component; component *MTPDS_GUI* has one connection with each of the *Age_Logic* and *Overweight_Logic* components; component *Age_Logic* has a connection with the *Personal_Database* component; component *Overweight_Logic* has a connection with the *Personal_Database* component.

11-5 Structure-Behavior Coalescence Diagram

We use a structure-behavior coalescence diagram (SBCD) to define the systems structure and systems behavior coexisting in the *Multi-Tier Personal Data System* as shown in Figure 11-13. SBCD is the fifth fundamental diagram to achieve structure-behavior coalescence. In the figure, interactions among the actor *Pupil* and the *MTPDS_GUI*, *Age_Logic*, *Overweight_Logic*, *Personal_Database* components shall draw forth the *AgeCalculation* and *OverweightCalculation* behaviors.

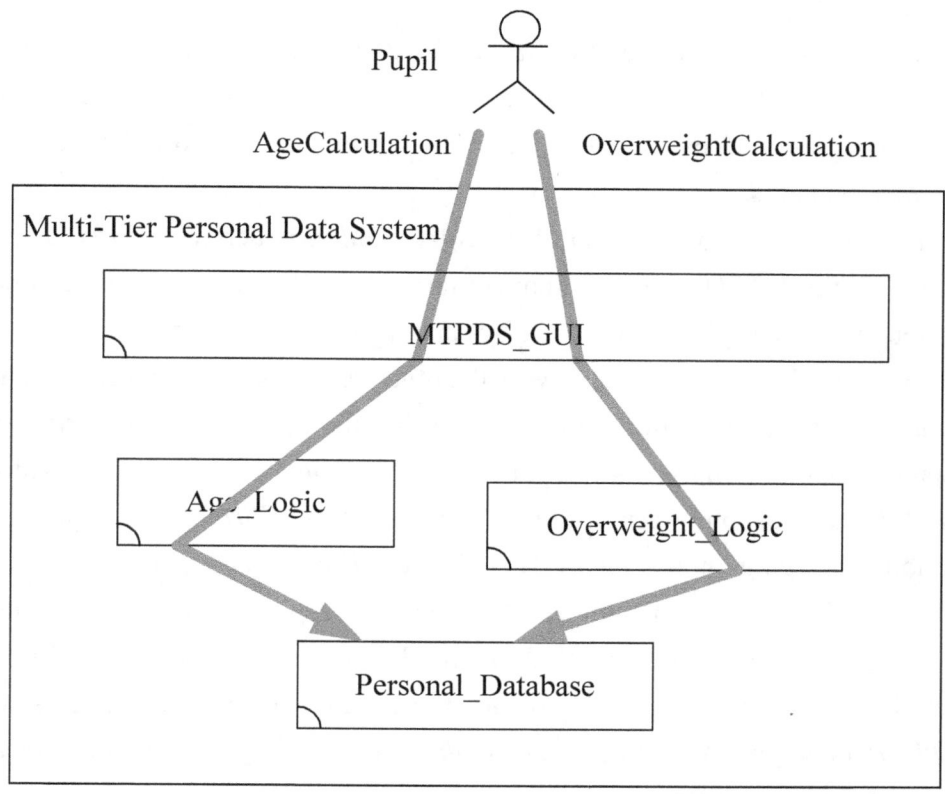

Figure 11-13   SBCD of the *Multi-Tier Personal Data System*

The overall behavior of a system is the aggregation of all its individual behaviors. For example, the overall behavior of the *Multi-Tier Personal Data System* includes the *AgeCalculation* and *OverweightCalculation* behaviors.

Be noticed that the *AgeCalculation* and *OverweightCalculation* behaviors are mutually independent of each other. They tend to be executed concurrently [Hoar85, Miln89, Miln99].

The major purpose of SBC architecture description language is to define an integrated whole of a system. In Figure 11-13, we not only see its systems structure, but also see at the same time its systems behavior in the SBCD of the *Multi-Tier Personal Data System.*

11-6 Interaction Flow Diagram

The overall behavior of the *Multi-Tier Personal Data System* includes two individual behaviors: *AgeCalculation* and *OverweightCalculation*. Each individual behavior is represented by an execution path. We use an IFD to define each one of these execution paths.

Figure 11-14 shows an IFD of the *AgeCalculation* behavior. First, actor *Pupil* interacts with the *MTPDS_GUI* component through the *Calculate_AgeClick* operation call interaction, carrying the *Social_Security_Number* input parameter. Next, component *MTPDS_GUI* interacts with the *AgeCalculation* component through the *Calculate_Age* operation call interaction, carrying the *Social_Security_Number* input parameter. Continuingly, component *Age_Logic* interacts with the *Personal_Database* component through the *Sql_DateOfBirth_Select* operation call interaction, carrying the *Social_Security_Number* input parameter and the *query_DateOfBirth* output parameter. Repeatedly, component *MTPDS_GUI* interacts with the *Age_Logic* component through the *Calculate_Age* operation return interaction, carrying the *Age* output parameter. Finally, actor *Pupil* interacts with the *MTPDS_GUI* component through the *Calculate_AgeClick* operation return interaction, carrying the *Age* output parameter.

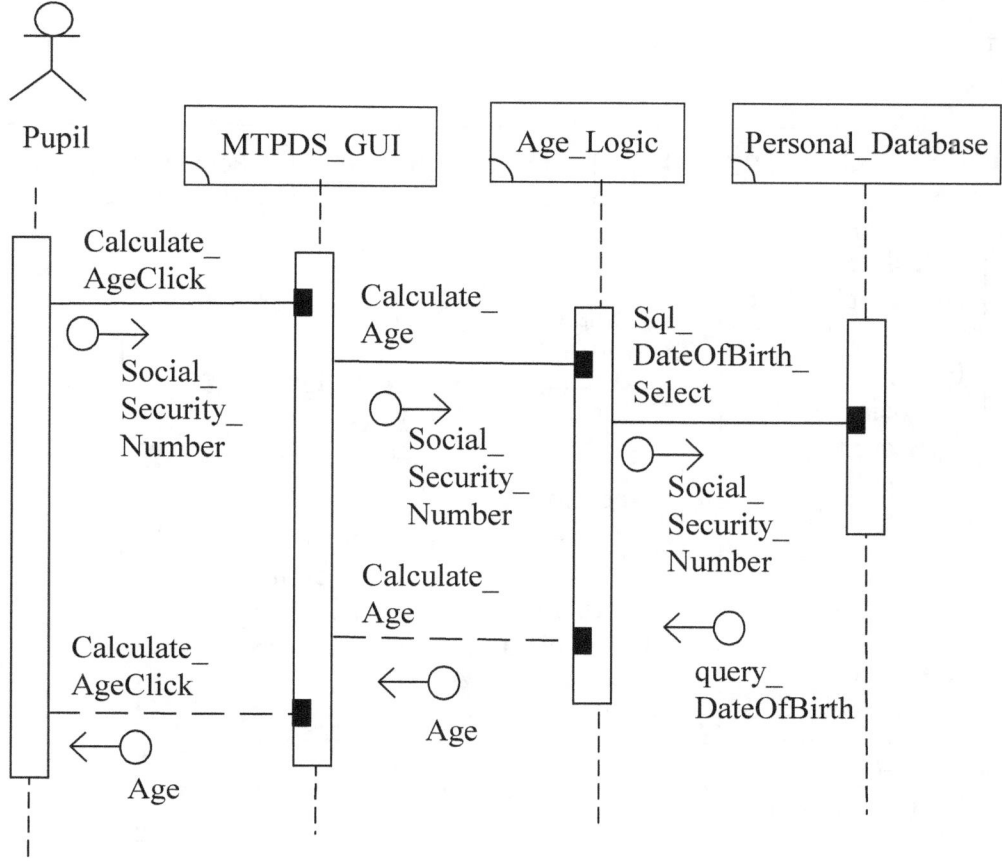

Figure 11-14   IFD of the *AgeCalculation* Behavior

Figure 11-15 shows an IFD of the *OverweightCalculation* behavior. First, actor *Pupil* interacts with the *MTPDS_GUI* component through the *Calculate_OverweightClick* operation call interaction, carrying the *Social_Security_Number* input parameter. Next, component *MTPDS_GUI* interacts with the *OverweightCalculation* component through the *Calculate_Overweight* operation call interaction, carrying the *Social_Security_Number* input parameter. Continuingly, component *Overweight_Logic* interacts with the *Personal_Database* component through the *Sql_SexHeightWeight_Select* operation call interaction, carrying the *Social_Security_Number* input parameter and the *query_SexHeightWeight* output parameter. Repeatedly, component *MTPDS_GUI* interacts with the *Overweight_Logic* component through the *Calculate_Overweight* operation return interaction, carrying the *Overweight* output parameter. Finally, actor *Pupil* interacts with the *MTPDS_GUI* component through the *Calculate_OverweightClick* operation return interaction, carrying the *Overweight* output parameter.

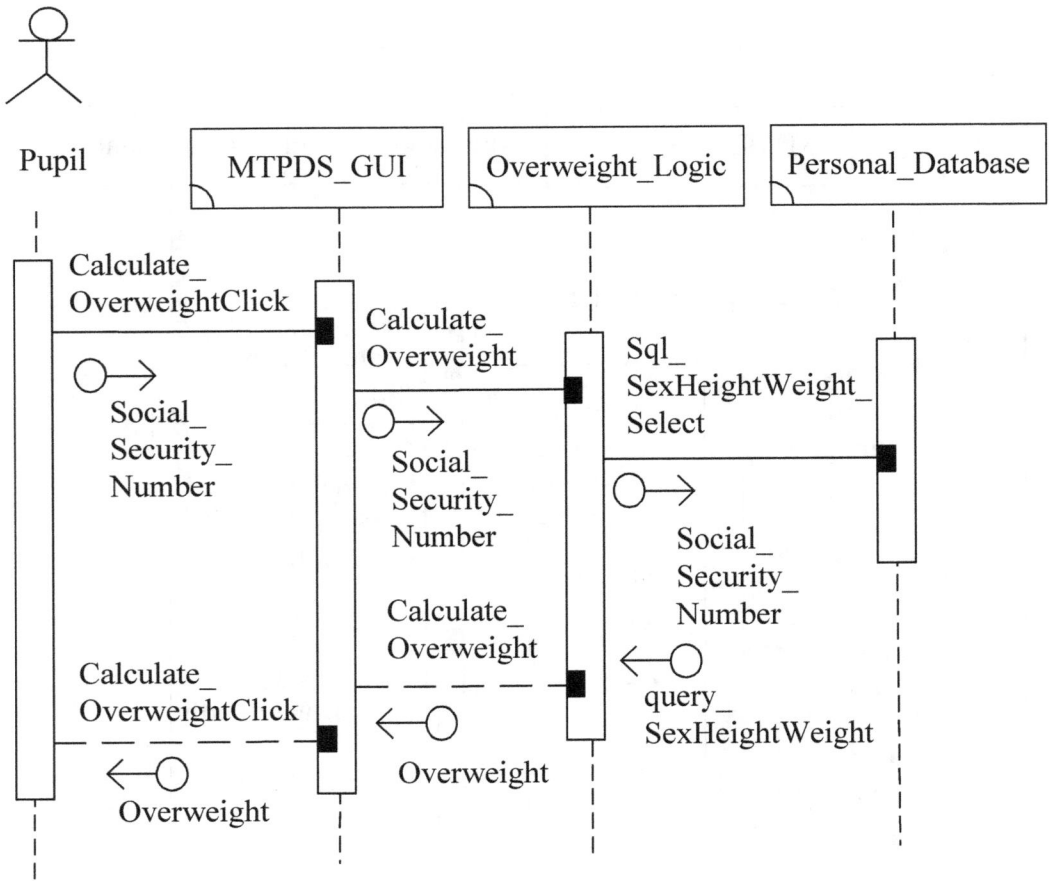

Figure 11-15   IFD of the *OverweightCalculation* Behavior

Chapter 12: Systems Thinking 2.0 Defining the Sale and Purchase System

This chapter demonstrates how to achieve systems thinking 2.0 (architectural thinking) defining the *sale and purchase system*, through the application of SBC architecture description language (SBC-ADL).

Sale and purchase system uses the *SalePurchase_GUI* component to provide two affairs. The first is the *Sale* affair which consists of the *SaleInput* and *SalePrint* behaviors, as shown in Figure 12-1.

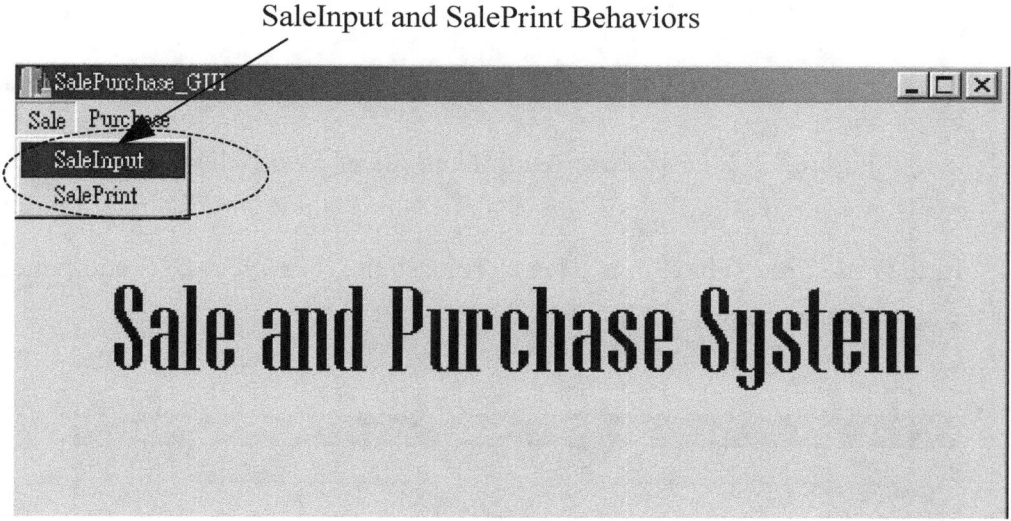

Figure 12-1   *Sale* Affair Consists of Two Behaviors

The second is the *Purchase* affair which consists of the *PurchaseInput* and *PurchasePrint* behaviors, as shown in Figure 12-2.

98

PurchaseInput and PurchasePrint Behaviors

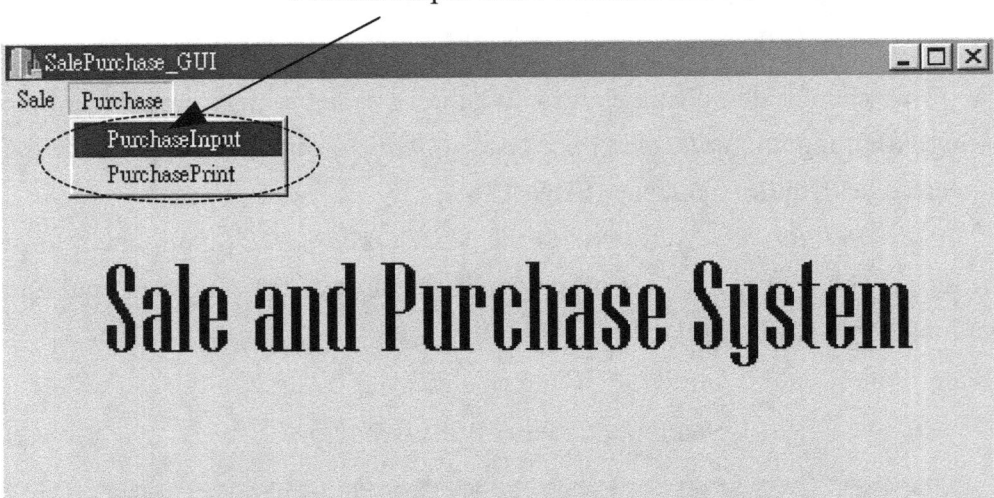

Figure 12-2   *Purchase* Affair Consists of Two Behaviors

In the *SaleInput* behavior, a sales clerk uses the *SaleInput_GUI* component to input the sale data, as shown in Figure 12-3.

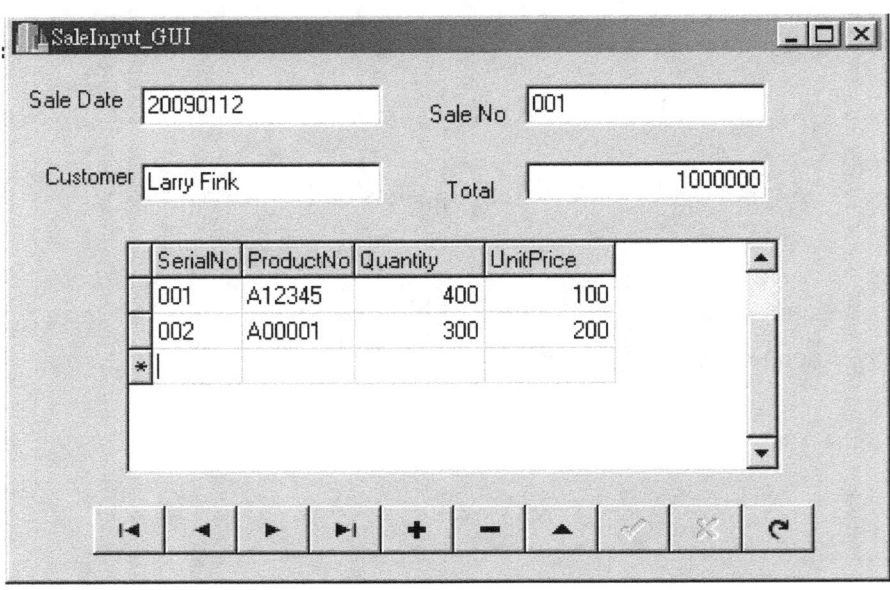

Figure 12-3   Input the Sale Data

In the *SalePrint* behavior, a sales clerk uses the *SalePrint_GUI* component to print out the sale data, as shown in Figure 12-4.

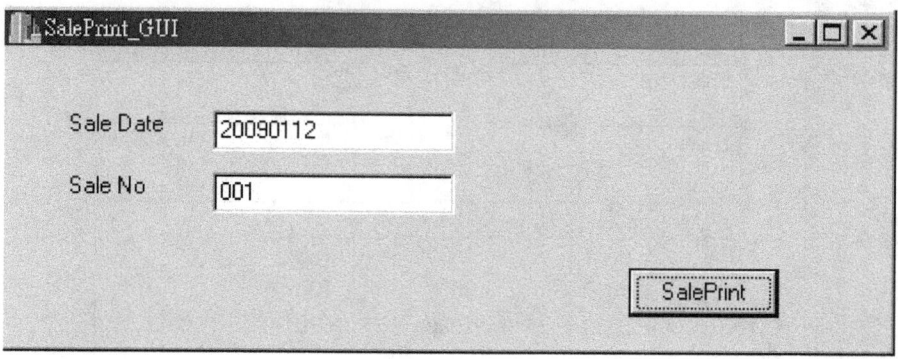

Figure 12-4   Print Out the Sale Data

In Figure 12-4, a sales clerk types the *Sale Date* and *Sale No* input data first. After pressing down the *SalePrint* button, the sales clerk then obtains the *Sale Data Report* output data as shown in Figure 12-5.

Sale Date : 20090112   Sale No : 001

Customer : Larry Fink

ProductNo	Quantity	UnitPrice
A12345	400	100
A00001	300	200

Total : 100,000

Figure 12-5   *Sale Data Report*

In the *PurchaseInput* behavior, a purchase clerk uses the PurchaseInput_GUI component to input the purchase data, as shown in Figure 12-6.

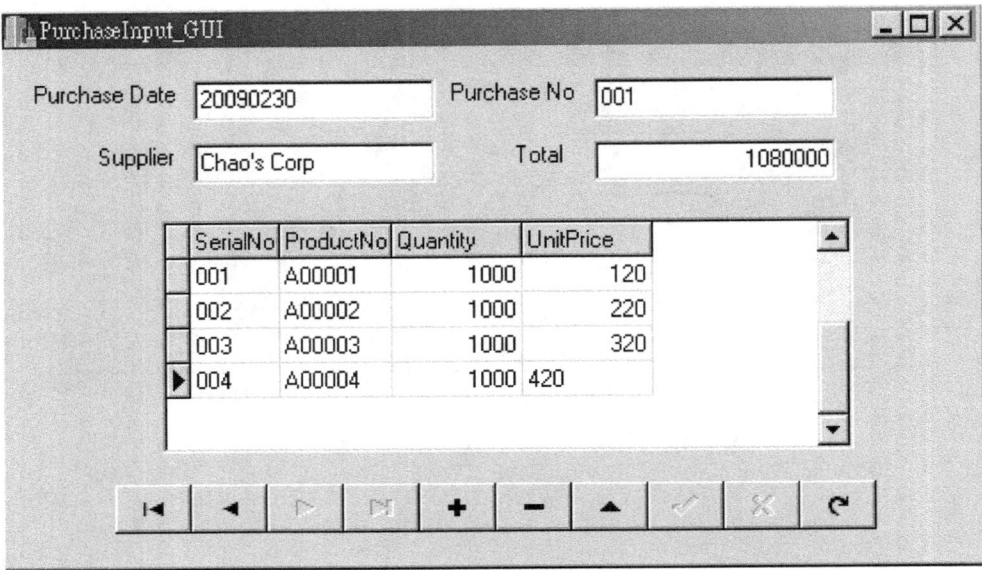

Figure 12-6   Input the Purchase Data

In the *PurchasePrint* behavior, a purchase clerk uses the *PurchasePrint_GUI* component to print out the purchase data, as shown in Figure 12-7.

Figure 12-7   Print Out the Purchase Data

In Figure 12-7, a purchase clerk types the *Purchase Date* and *Purchase No* input data first. After pressing down the *PurchasePrint* button, the purchase clerk then obtains the *Purchase Data Report* output data as shown in Figure 12-8.

Purchase Date : 20090230   Purchase No : 001

Supplier : Chao's Corp

ProductNo	Quantity	UnitPrice
A00001	1000	120
A00002	1000	220
A00003	1000	320
A00004	1000	420

Total : 1,080,000

Figure 12-8.   *Purchase Data Report*

Using the SBC architecture description language, we shall go through: a) architecture hierarchy diagram, b) framework diagram, c) component operation diagram, d) component connection diagram, e) structure-behavior coalescence diagram and f) interaction flow diagram, to accomplish systems thinking 2.0 (architectural thinking) defining the *sale and purchase system.*

12-1 Architecture Hierarchy Diagram

We use an architecture hierarchy diagram (AHD) to define the multi-level composition and decomposition of the *sale and purchase system*. AHD is the first fundamental diagram to achieve structure-behavior coalescence. As shown in Figure 12-9, *Sale and Purchase System* is composed of *SalePurchase_GUI, SaleInput_GUI, SalePrint_GUI, PurchaseInput_GUI, PurchasePrint_GUI* and *S_Subsystem_1*; *S_Subsystem_1* is composed of *SalePurchase_Database.*

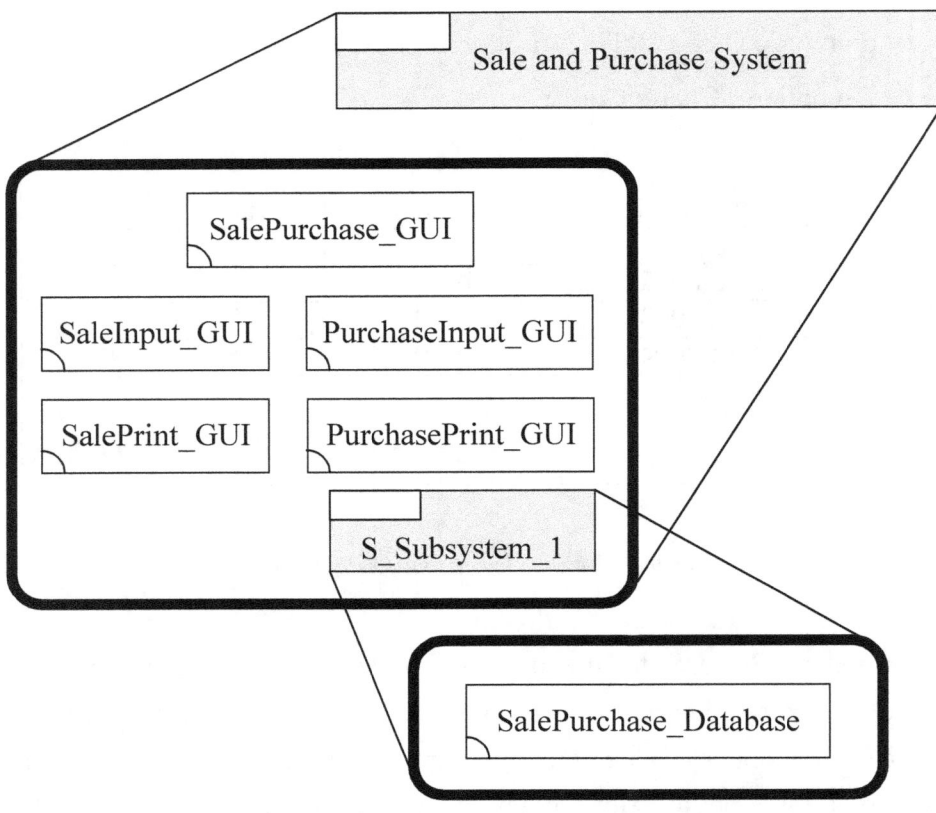

Figure 12-9   AHD of the *Sale and Purchase System*

In Figure 12-9, *Sale and Purchase System* and *S_Subsystem_1* are aggregated systems while *SalePurchase_GUI*, *SaleInput_GUI*, *SalePrint_GUI*, *PurchaseInput_GUI*, *PurchasePrint_GUI* and *SalePurchase_Database* are non-aggregated systems.

12-2 Framework Diagram

We use a framework diagram (FD) to define the multi-layer composition and decomposition of the *sale and purchase system* as shown in Figure 12-10. FD is the second fundamental diagram to achieve structure-behavior coalescence. In the figure, *Application_Layer* contains the *SalePurchase_GUI*, *SaleInput_GUI*, *SalePrint_GUI*, *PurchaseInput_GUI* and *PurchasePrint_GUI* components; *Data_Layer* contains the *SalePurchase_Database* component.

Figure 12-10   FD of the *Sale and Purchase System*

12-3 Component Operation Diagram

We use a component operation diagram (COD) to define the operations of all components of the *sale and purchase system* as shown in Figure 12-11. COD is the third fundamental diagram to achieve structure-behavior coalescence. In the figure, component *SalePurchase_GUI* has four operations: *SaleInputClick*, *SalePrintClick*, *PurchaseInputClick* and *PurchasePrintClick*; component *SaleInput_GUI* has two operations: *ShowModal* and *SaleDataInput*; component *SalePrint_GUI* has two operations: *ShowModal* and *SalePrintButtonClick*; component *PurchaseInput_GUI* has two operations: *ShowModal* and *PurchaseDataInput*; component *PurchaseInput_GUI* has two operations: *ShowModal* and *PurchasePrintButtonClick*; component *SalePurchase_Database* has four operations: *Sql_s_insert*, *Sql_s_select*, *Sql_p_insert* and *Sql_p_select*.

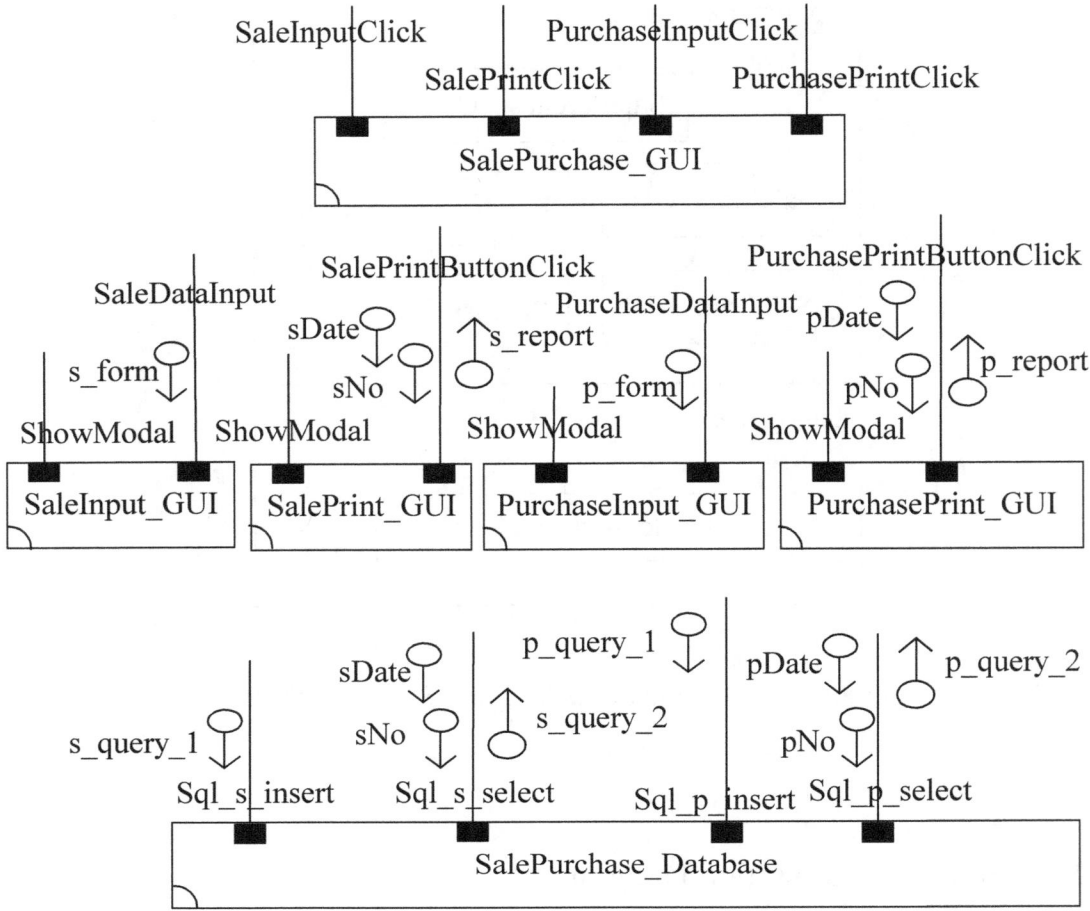

Figure 12-11   COD of the *Sale and Purchase System*

The operation formula of *SaleDataInput* is *SaleDataInput(In s_form)*. The operation formula of *SalePrintButtonClick* is *SalePrintButtonClick(In sDate, sNo; Out s_report)*. The operation formula of *PurchaseDataInput* is *PurchaseDataInput(In p_form)*. The operation formula of *PurchasePrintButtonClick* is *PurchasePrintButtonClick(In pDate, pNo; Out p_report)*. The operation formula of *Sql_s_insert* is *Sql_s_insert(In s_query_1)*. The operation formula of *Sql_s_select* is *Sql_s_select(In sDate, sNo; Out s_query_2)*. The operation formula of *Sql_p_insert* is *Sql_p_insert(In p_query_1)*. The operation formula of *Sql_p_select* is *Sql_p_select(In pDate, pNo; Out p_query_2)*.

Figure 12-12 shows the primitive data type specification of the *sDate*, *sNo*, *pDate* and *pNo* input parameters.

Parameter	Data Type	Instances
sDate	Text	20100517, 20100612
sNo	Text	001, 002
pDate	Text	20110317, 20110412
pNo	Text	003, 004

Figure 12-12   Primitive Data Type Specification

Figure 12-13 shows the composite data type specification of the *s_form* input parameter occurring in the *SaleDataInput(In s_form)* operation formula.

Parameter	*s_form*

Data Type	TABLE of
	    Sale Date : Text
	    Customer : Text
	    ProductNo : Text
	    Quantity : Integer
	    UnitPrice : Real
	    Total : Real
	End TABLE ;

Instances	**Sale Input Form**
	
	    Sale Date: 2010/05/17
	
	Customer : __Larry Fink__
	
	ProductNo Quantity Unit Price
	
	____A12345_____400_____100.00____
	____A00001_____300_____200.00____
	
	    Merchandise Total : 100,000.00

Figure 12-13   Composite Data Type Specification

Figure 12-14 shows the composite data type specification of the *s_report* output parameter occurring in the *SalePrintButtonClick(In sDate, sNo; Out s_report)* operation formula.

Parameter	*s_report*

Data Type	TABLE of
	    Sale Date : Text
	    Sale No : Text
	    Customer : Text
	    ProductNo : Text
	    Quantity : Integer
	    UnitPrice : Real
	    Total : Real
	End TABLE ;

Instances	Sale Date : 20100517 Sale No : 001 Customer : Larry Fink ProductNo / Quantity / UnitPrice A12345 / 400 / 100.00 A00001 / 300 / 200.00 Total : 100,000.00

Figure 12-14   Composite Data Type Specification

Figure 12-15 shows the composite data type specification of the *p_form* input parameter occurring in the *PurchaseDataInput(In p_form)* operation formula.

Parameter	*p_form*
Data Type	TABLE of 　Purchase Date : Text 　Supplier :　Text 　ProductNo : Text 　Quantity : Integer 　UnitPrice : Real 　Total : Real End TABLE ;
Instances	**Purchase　Input　Form** 　　　　　　　　　　Purchase Date:__2010/06/12 Supplier : _Chao's Corp._ 　ProductNo　　　　Quantity　　　　Unit Price __A00001_____1000_____120.00____ __A00002_____2000_____220.00____ __A00003_____3000_____320.00____ __A00004_____4000_____420.00____ 　　　　　　Merchandise Total : 1,080,000.00

Figure 12-15　　Composite Data Type Specification

Figure 12-16 shows the composite data type specification of the *p_report* output parameter occurring in the *PurchasePrintButtonClick(In pDate, pNo; Out p_report)* operation formula.

Parameter	*p_report*

Data Type	TABLE of
	  Purchase Date : Text
	  Purchase No : Text
	  Supplier : Text
	  ProductNo : Text
	  Quantity : Integer
	  UnitPrice : Real
	  Total : Real End TABLE ;

Instances	Purchase Date : 20100612 Purchase No : 001 Supplier : Chao's Corp ProductNo | Quantity | UnitPrice A00001 | 1000 | 120.00 A00002 | 1000 | 220.00 A00003 | 1000 | 320.00 A00004 | 1000 | 420.00 Total : 1,080,000.00

Figure 12-16   Composite Data Type Specification

Figure 12-17 shows the composite data type specification of the *s_query_1* input parameter occurring in the *Sql_s_insert(In s_query_1)* operation formula.

Parameter	*s_query_1*

Data Type	TABLE of
	  Sale Date : Text
	  Sale No : Text
	  Customer : Text
	  ProductNo : Text
	  Quantity : Integer
	  UnitPrice : Real
	  Total : Real End TABLE ;

Instances	Sale Date | Sale No | Customer | Total 20090112 | 001 | Larry Fink | 100,000.00 ProductNo | Quantity | UnitPrice A12345 | 400 | 100.00 A00001 | 300 | 200.00

Figure 12-17   Composite Data Type Specification

Figure 12-18 shows the composite data type specification of the *s_query_2* output parameter occurring in the *Sql_s_select(Out s_query_2)* operation formula.

Parameter	*s_query_2*		

Data Type	TABLE of
	  Sale Date : Text
	  Sale No : Text
	  Customer : Text
	  ProductNo : Text
	  Quantity : Integer
	  UnitPrice : Real
	  Total : Real
	End TABLE ;		

Instances			

Sale Date	Sale No	Customer	Total
20090112	001	Larry Fink	100,000.00

ProductNo	Quantity	UnitPrice
A12345	400	100.00
A00001	300	200.00

Figure 12-18   Composite Data Type Specification

Figure 12-19 shows the composite data type specification of the *p_query_1* input parameter occurring in the *Sql_p_insert(In p_query_1)* operation formula.

Parameter	*p_query_1*

Data Type	TABLE of
	    Purchase Date : Text
	    Purchase No : Text
	    Supplier : Text
	    ProductNo : Text
	    Quantity : Integer
	    UnitPrice : Real
	    Total : Real
	End TABLE ;

Instances	

Purchase Date	Purchase No	Supplier	Total
20090230	001	Chao's Corp	1,080,000.00

ProductNo	Quantity	UnitPrice
A00001	1000	120.00
A00002	1000	220.00
A00003	1000	320.00
A00004	1000	420.00

Figure 12-19   Composite Data Type Specification

Figure 12-20 shows the composite data type specification of the *p_query_2* output parameter occurring in the *Sql_p_select(In p_query_2)* operation formula.

Parameter	*p_query_2*

Data Type	TABLE of
	   Purchase Date : Text
	   Purchase No : Text
	   Supplier : Text
	   ProductNo : Text
	   Quantity : Integer
	   UnitPrice : Real
	   Total : Real
	End TABLE ;

Instances	

Purchase Date	Purchase No	Supplier	Total
20090230	001	Chao's Corp	1,080,000.00

ProductNo	Quantity	UnitPrice
A00001	1000	120.00
A00002	1000	220.00
A00003	1000	320.00
A00004	1000	420.00

Figure 12-20   Composite Data Type Specification

12-4 Component Connection Diagram

We use a component connection diagram (CCD) to define the connections among components and actors of the *sale and purchase system* as shown in Figure 12-21. CCD is the fourth fundamental diagram to achieve structure-behavior coalescence.

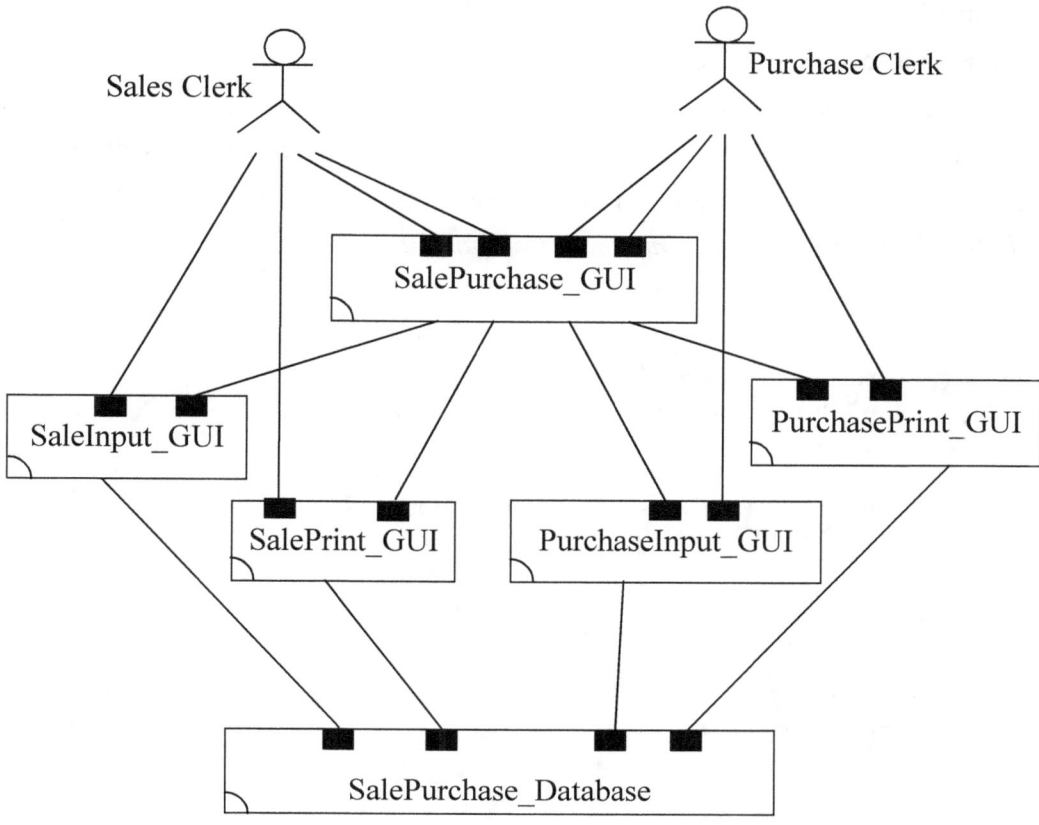

Figure 12-21   CCD of the *Sale and Purchase System*

In Figure 12-21, actor *Sales Clerk* has two connections with the *SalePurchase_GUI* components; actor *Sales Clerk* has one connection with each one of the *SaleInput_GUI* and *SalePrint_GUI* components; actor *Purchase Clerk* has two connections with the *SalePurchase_GUI* components; actor *Purchase Clerk* has one connection with each one of the *PurchaseInput_GUI* and *PurchasePrint_GUI* components; component *SalePurchase_GUI* has one connection with each one of the *SaleInput_GUI*, *PurchasePrint_GUI*, *SaleInput_GUI* and *PurchaseInput_GUI* component. Each one of the *SaleInput_GUI*, *SalePrint_GUI*, *PurchaseInput_GUI* and *PurchasePrint_GUI* components has one connection with the *SalePurchase_Database* component.

12-5 Structure-Behavior Coalescence Diagram

We use a structure-behavior coalescence diagram (SBCD) to define the systems structure and systems behavior coexisting in the *sale and purchase system* as shown in Figure 12-22. SBCD is the fifth fundamental diagram to achieve structure-behavior coalescence.

In the figure, interactions among the *Sales_Clerk*, *Purchase_Clerk* actors and the *SalePurchase_GUI*, *SaleInput_GUI*, *SalePrint_GUI*, *PurchaseInput_GUI*, *PurchasePrint_GUI*, *SalePurchase_Database* components shall draw forth the *SaleInput*, *SalePrint*, *PurchaseInput* and *PurchasePrint* behaviors.

Figure 12-22   SBCD of the *Sale and Purchase System*

The overall behavior of a system is the aggregation of all its individual behaviors. For example, the overall behavior of the *sale and purchase system* includes the *SaleInput*, *SalePrint*, *PurchaseInput* and *PurchasePrint* behaviors.

Be noticed that the *SaleInput*, *SalePrint*, *PurchaseInput* and *PurchasePrint* behaviors are mutually independent of each other. They tend to be executed concurrently [Hoar85, Miln89, Miln99].

The major purpose of SBC architecture description language is to define an integrated whole of a system. In Figure 12-22, we not only see its structure, but also see at the same time its behavior in the *sale and purchase system*'s SBCD.

12-6 Interaction Flow Diagram

The overall behavior of the *sale and purchase system* includes four individual behaviors: *SaleInput*, *SalePrint*, *PurchaseInput* and *PurchasePrint*. Each individual behavior is represented by an execution path. We use an IFD to define each one of these execution paths.

Figure 12-23 shows an IFD of the *SaleInput* behavior. First, actor *Sales_Clerk* interacts with the *SalePurchase_GUI* component through the *SaleInputClick* operation call interaction. Next, component *SalePurchase_GUI* interacts with the *SaleInput_GUI* component through the *ShowModal* operation call interaction. Continuingly, actor *Sales Clerk* interacts with the *SaleInput_GUI* component through the *SaleDataInput* operation call interaction, carrying the *s_form* input parameter. Finally, component *SaleInput_GUI* interacts with the *SalePurchase_Database* component through the *Sql_s_insert* operation call interaction, carrying the *s_query_1* input parameter.

Figure 12-23   IFD of the *SaleInput* Behavior

Figure 12-24 shows an IFD of the *SalePrint* behavior. First, actor *Sales_Clerk* interacts with the *SalePurchase_GUI* component through the *SalePrintClick* operation call interaction. Next, component *SalePurchase_GUI* interacts with the *SalePrint_GUI* component through the *ShowModal* operation call interaction. Continuingly, actor *Sales Clerk* interacts with the *SalePrint_GUI* component through the *SalePrintButtonClick* operation call interaction, carrying the *sDate* and *sNo* input parameters. Continuingly, component *SalePrint_GUI* interacts with the *SalePurchase_Database* component through the *Sql_s_select* operation call interaction, carrying the *sDate* and *sNo* input parameters and the *s_query_2* output parameter. Finally, actor *Sales Clerk* interacts with the *SalePrint_GUI* component through the *SalePrintButtonClick* operation return interaction, carrying the *s_report* output parameter.

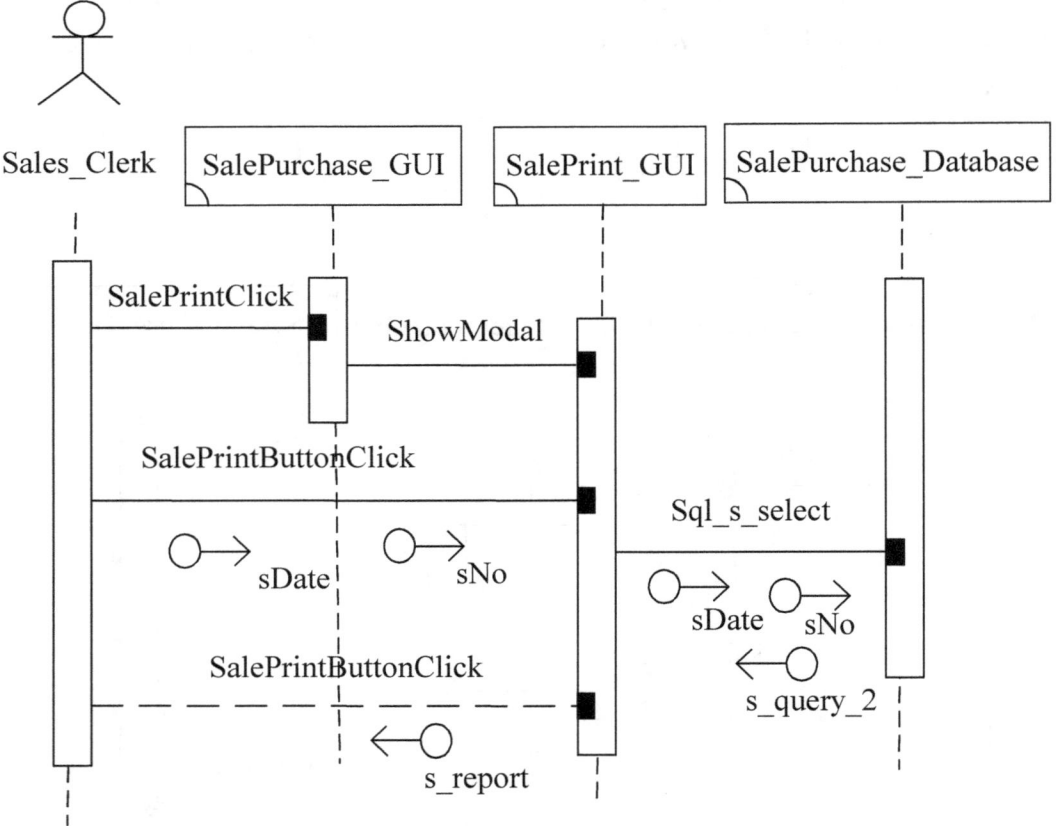

Figure 12-24   IFD of the *SalePrint* Behavior

Figure 12-25 shows an IFD of the *PurchaseInput* behavior. First, actor *Purchase_Clerk* interacts with the *SalePurchase_GUI* component through the *PurchaseInputClick* operation call interaction. Next, component *SalePurchase_GUI* interacts with the *PurchaseInput_GUI* component through the *ShowModal* operation call interaction. Continuingly, actor *Purchase Clerk* interacts with the *PurchaseInput_GUI* component through the *PurchaseDataInput* operation call interaction, carrying the *p_form* input parameter. Finally, component *PurchaseInput_GUI* interacts with the *SalePurchase_Database* component through the *Sql_p_insert* operation call interaction, carrying the *p_query_1* input parameter.

Figure 12-25   IFD of the *PurchaseInput* Behavior

Figure 12-26 shows an IFD of the *PurchasePrint* behavior. First, actor *Purchase_Clerk* interacts with the *SalePurchase_GUI* component through the *PurchasePrintClick* operation call interaction. Next, component *SalePurchase_GUI* interacts with the *PurchasePrint_GUI* component through the *ShowModal* operation call interaction. Continuingly, actor *Purchase Clerk* interacts with the *PurchasePrint_GUI* component through the *PurchasePrintButtonClick* operation call interaction, carrying the *sDate* and *sNo* input parameters. Continuingly, component *PurchasePrint_GUI* interacts with the *SalePurchase_Database* component through the *Sql_p_select* operation call interaction, carrying the *pDate* and *pNo* input parameters and the *p_query_2*output parameter. Finally, actor *Purchase Clerk* interacts with the *PurchasePrint_GUI* component through the *PurchasePrintButtonClick* operation return interaction, carrying the *p_report* output parameter.

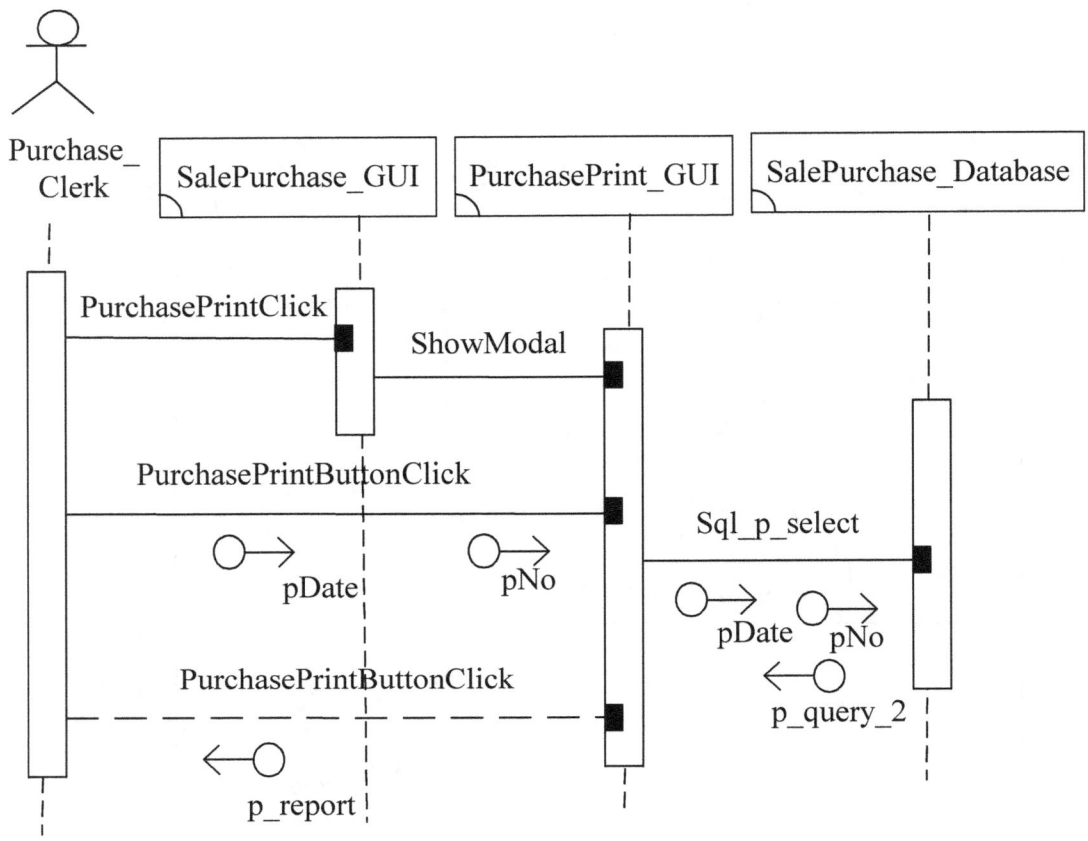

Figure 12-26   IFD of the *PurchasePrint* Behavior

Chapter 13: Systems Thinking 2.0 Defining the Web Service Arithmetic System

This chapter demonstrates how to achieve systems thinking 2.0 (architectural thinking) defining the *web service arithmetic system*, through the application of SBC architecture description language (SBC-ADL). After the system development is completed, the *web service arithmetic system* shall appear on a web multi-tier platform [Sebe12] as shown in Figure 13-1.

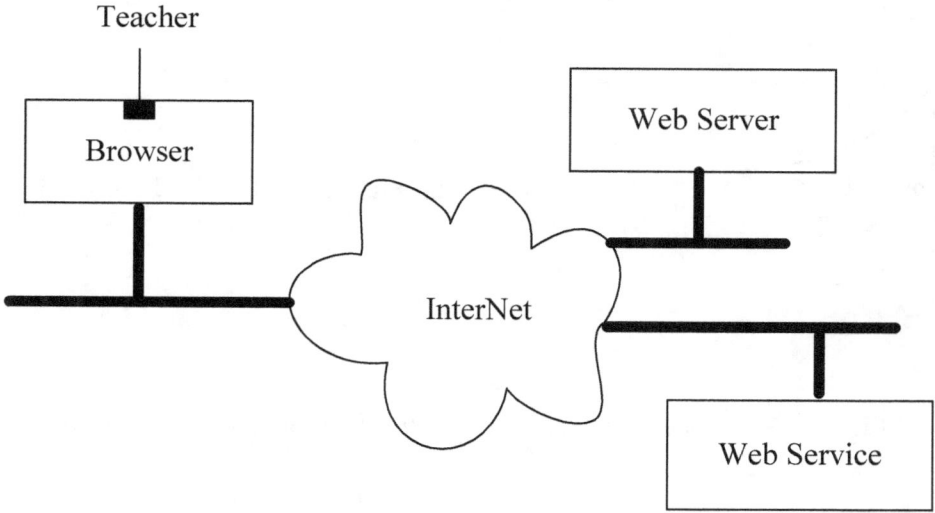

Figure 13-1 *Web Service Arithmetic System* on a Web Multi-Tier Platform

The functionality of the *Web Service Arithmetic System* is to provide a web browser for the *Teacher* actor to input the *P*, *Q*, *R*, *S* and *T* values as shown in Figure 13-2.

120

Input the *P, Q, R, S, T* values

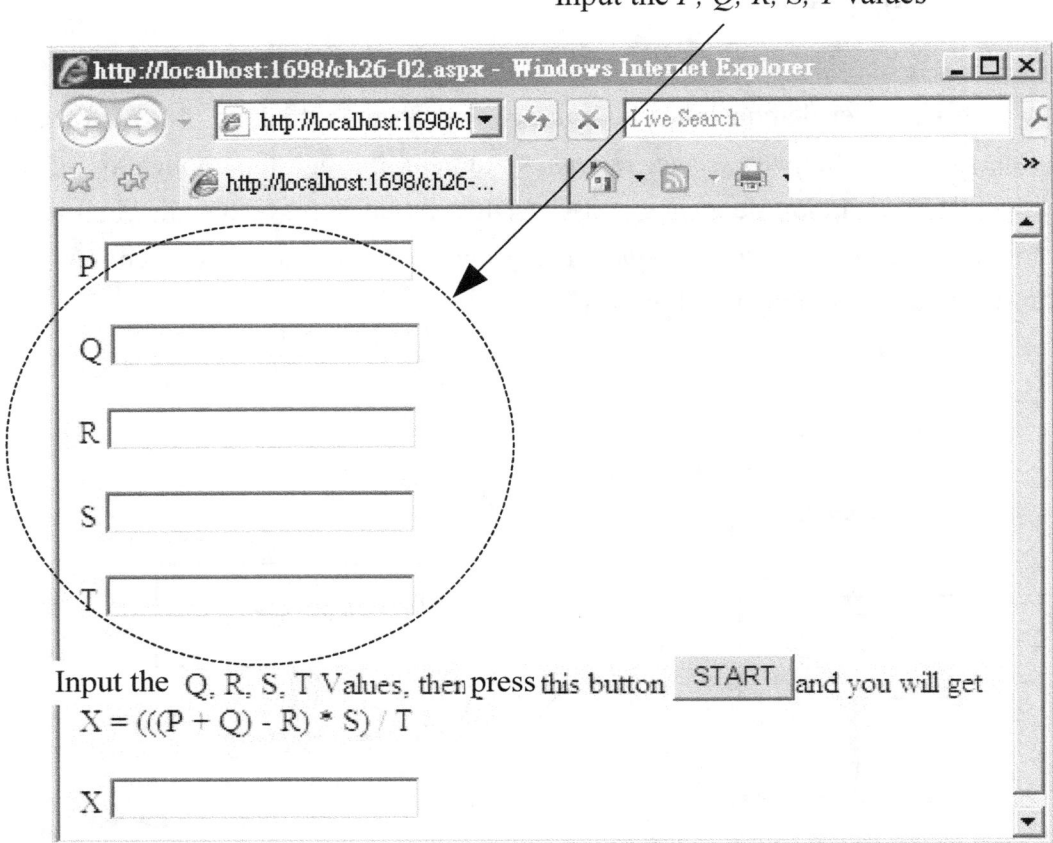

Figure 13-2　Web Browser for the *Pupil* Actor to Input the Values

When the *START* button is pressed, the *Web Service Arithmetic System* calculates the (((P+Q)-R) *S)/T value and presents the result on the *X* output box, as shown in Figure 13-3.

$$23 = (((20+2)-8)*5)/3$$

Figure 13-3   Result after Pressing Down the *START* Button

Using the SBC architecture description language, we shall go through: a) architecture hierarchy diagram, b) framework diagram, c) component operation diagram, d) component connection diagram, e) structure-behavior coalescence diagram and f) interaction flow diagram, to accomplish systems thinking 2.0 (architectural thinking) defining the *web service arithmetic system.*

13-1 Architecture Hierarchy Diagram

We use an architecture hierarchy diagram (AHD) to define the multi-level composition and decomposition of the *web service arithmetic system*. AHD is the first fundamental diagram to achieve structure-behavior coalescence. As shown in Figure 13-4, *Web Service Arithmetic System* is composed of *WebServerArithmetic_GUI* and *W_Subsystem_1*; *W_Subsystem_1* is composed of *WebServiceArithmetic.*

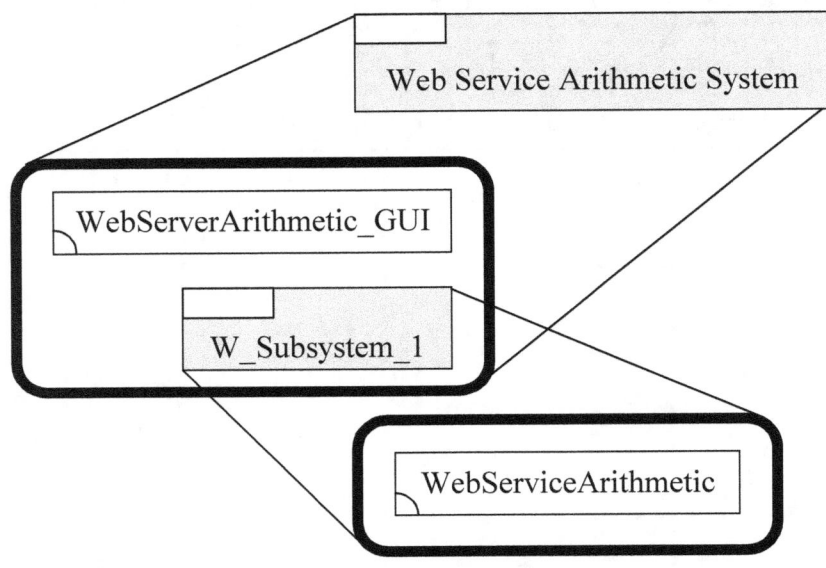

Figure 13-4   AHD of the *Web Service Arithmetic System*

In Figure 13-4, *Web Service Arithmetic System* and *W_Subsystem_1* are aggregated systems while *WebServerArithmetic_GUI* and *WebServiceArithmetic* are non-aggregated systems.

13-2 Framework Diagram

We use a framework diagram (FD) to define the multi-layer composition and decomposition of the *web service arithmetic system* as shown in Figure 13-5. FD is the second fundamental diagram to achieve structure-behavior coalescence. In the figure, *Application_Layer* contains the *WebServerArithmetic_GUI* component; *Logic_Layer* contains the *WebServiceArithmetic* component.

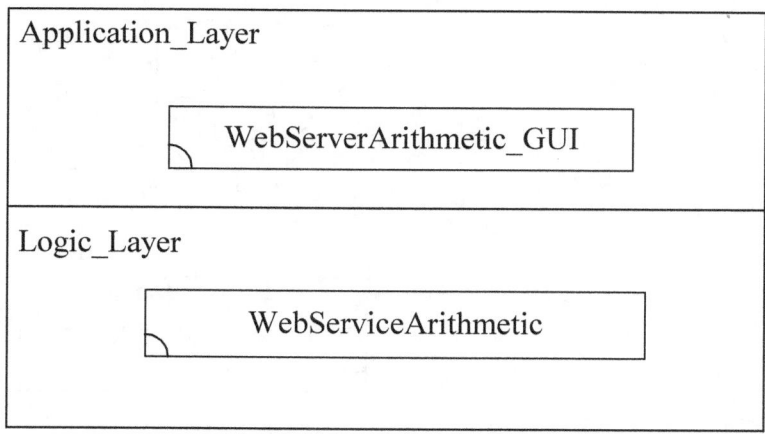

Figure 13-5   FD of the *Web Service Arithmetic System*

13-3 Component Operation Diagram

We use a component operation diagram (COD) to define the operations of all components of the *web service arithmetic system* as shown in Figure 13-6. COD is the third fundamental diagram to achieve structure-behavior coalescence. In the figure, component *WebServerArithmetic_GUI* has one operation: *START_Click*; component *WebServiceArithmetic* has four operations: *Add*, *Subtract*, *Multiply* and *Divide*.

124

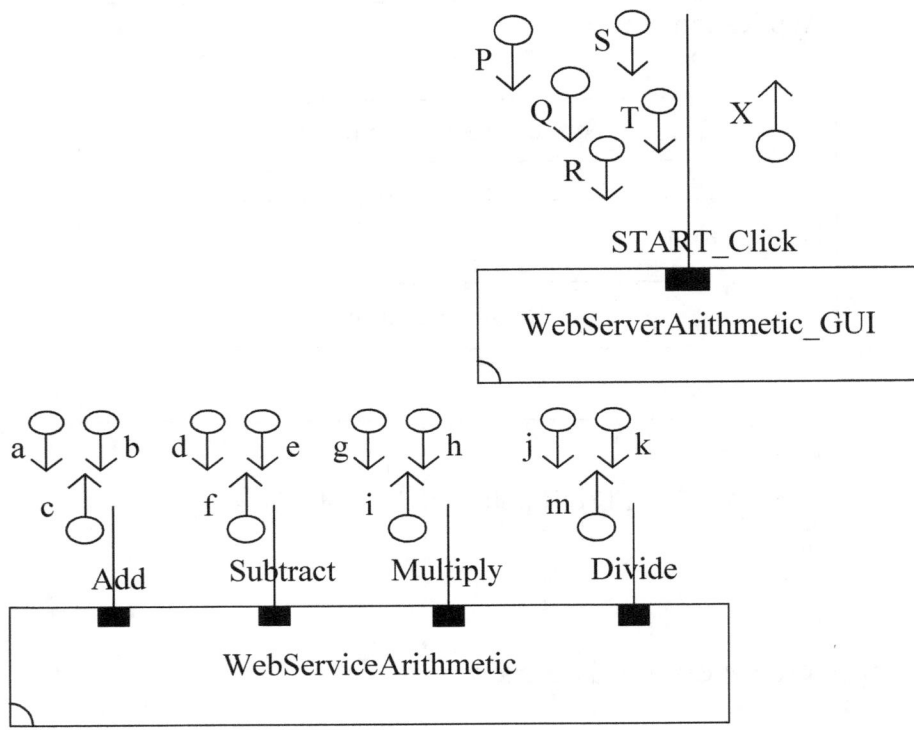

Figure 13-6   COD of the *Web Service Arithmetic System*

The operation formula of *START_Click* is *START_Click(In P, Q, R, S, T; Out X)*. The operation formula of *Add* is *Add(In a, b; Out c)*. The operation formula of *Subtract* is *Subtract(In d, e; Out f)*. The operation formula of *Multiply* is *Multiply(In g, h; Out i)*. The operation formula of *Divide* is *Divide(In j, k; Out m)*.

Figure 13-7 shows the primitive data type specification of the *P, Q, R, S, T, a, b, d, e, g, h, j, k* input parameter and the *X, c, f, i, m* output parameters.

Parameter	Data Type	Instances
P	Integer	20, 7
Q	Integer	2, 6
R	Integer	8, 5
S	Integer	5, 4
T	Integer	3, 3
X	Integer	23, 11
a	Integer	20, 7
b	Integer	2, 6
c	Integer	22, 13
d	Integer	22, 13
e	Integer	8, 5
f	Integer	14, 8
g	Integer	14, 8
h	Integer	5, 4
i	Integer	70, 32
j	Integer	70, 32
k	Integer	3, 3
m	Integer	23, 11

Figure 13-7   Primitive Data Type Specification

13-4 Component Connection Diagram

We use a component connection diagram (CCD) to define the connections among components and actors of the *web service arithmetic system* as shown in

Figure 13-8. CCD is the fourth fundamental diagram to achieve structure-behavior coalescence.

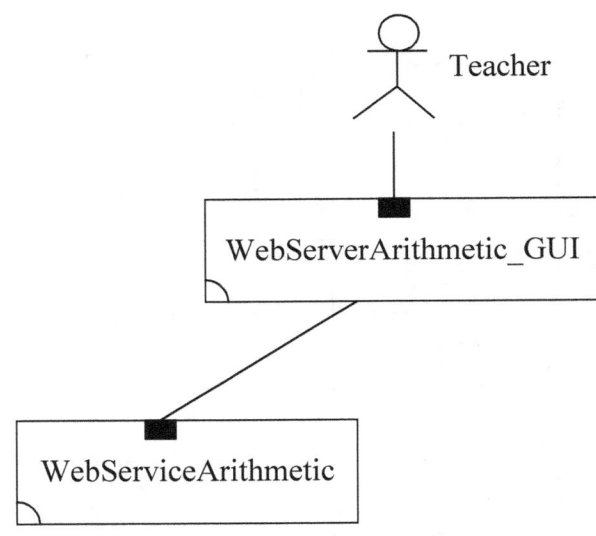

Figure 13-8   CCD of the *Web Service Arithmetic System*

In Figure 13-8, actor *Teacher* has one connection with the *WebServerArithmetic_GUI* component; component *WebServerArithmetic_GUI* has one connection with the *WebServiceArithmetic* component.

13-5 Structure-Behavior Coalescence Diagram

We use a structure-behavior coalescence diagram (SBCD) to define the systems structure and systems behavior coexisting in the *web service arithmetic system* as shown in Figure 13-9. SBCD is the fifth fundamental diagram to achieve structure-behavior coalescence. In the figure, interactions among the *Teacher* actor and the *WebServerArithmetic_GUI, WebServiceArithmetic* components shall draw forth the *Calculating (((P+Q)-R) *S)/T Value* behavior.

Figure 13-9   SBCD of the *Web Service Arithmetic System*

The overall behavior of a system is the aggregation of all its individual behaviors. For example, the overall behavior of the *web service arithmetic system* includes the *Calculating (((P+Q)-R) *S)/T Value* behavior.

The major purpose of SBC architecture description language is to define an integrated whole of a system. In Figure 13-9, we not only see its structure, but also see at the same time its behavior in the *web service arithmetic system*'s SBCD.

13-6 Interaction Flow Diagram

The overall behavior of the *web service arithmetic system* includes one individual behavior: *Calculating (((P+Q)-R) *S)/T Value*. Each individual behavior is represented by an execution path. We use an IFD to define each one of these execution paths.

Figure 13-10 shows an IFD of the *Calculating (((P+Q)-R) *S)/T Value* behavior. First, actor *Teacher* interacts with the *WebServerArithmetic_GUI* component through the *START_Click* operation call interaction, carrying the *P*, *Q*, R, *S* and *T* input parameters. Next, component *WebServerArithmetic_GUI* interacts with the *WebServiceArithmetic* component through the *Add* operation call interaction, carrying the *a* and *b* input parameters. Continuingly, component *WebServerArithmetic_GUI* interacts again with the *WebServiceArithmetic* component through the *Add* operation return interaction, carrying the *c* output parameter. Repeatedly, component *WebServerArithmetic_GUI* interacts with the *WebServiceArithmetic* component through the *Subtract* operation call interaction, carrying the *d* and *e* input parameters. Continuingly, component *WebServerArithmetic_GUI* interacts with the *WebServiceArithmetic* component through the *Subtract* operation return interaction, carrying the *f* output parameter. Repeatedly, component *WebServerArithmetic_GUI* interacts with the *WebServerArithmetic* component through the *Multiply* operation call interaction, carrying the *g* and *h* input parameters. Continuingly, component *WebServerArithmetic_GUI* interacts with the *WebServerArithmetic* component through the *Multiply* operation return interaction, carrying the *i* output parameter. Repeatedly, component *WebServerArithmetic_GUI* interacts with the *WebServerArithmetic* component through the *Divide* operation call interaction, carrying the *j* and *k* input parameters. Continuingly, component *WebServerArithmetic_GUI* interacts again with the *WebServerArithmetic* component through the *Divide* operation return interaction, carrying the *m* output parameter. Finally, actor *Teacher* interacts with the *WebServerArithmetic_GUI* component through the *START_Click* operation return interaction, carrying the *X* output parameter.

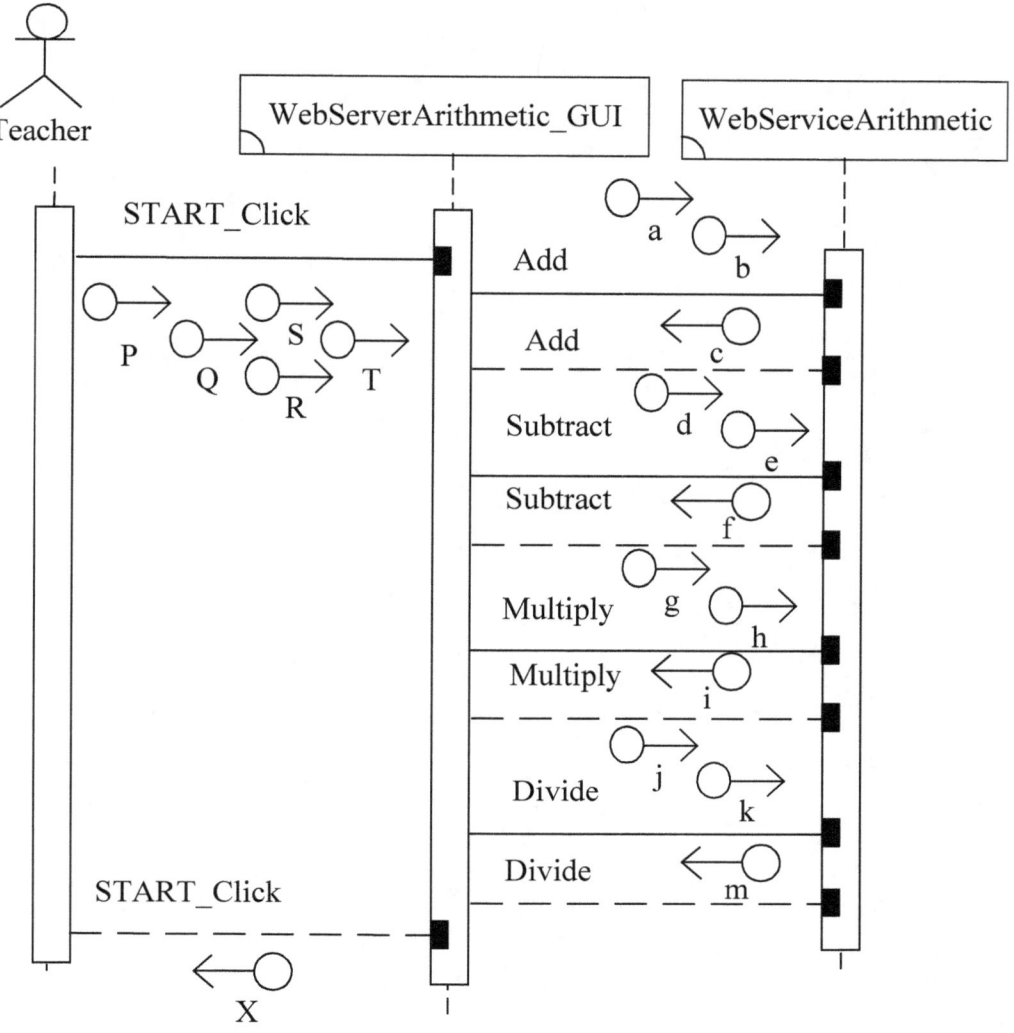

Figure 13-10   IFD of the *Calculating (((P+Q)-R) *S)/T Value* Behavior

Chapter 14: Systems Thinking 2.0 Defining the Web Service Extranet System

This chapter demonstrates how to achieve systems thinking 2.0 (architectural thinking) defining the *web service extranet system*, through the application of SBC architecture description language (SBC-ADL). After the system development is completed, the *web service extranet system* shall appear on a web multi-tier platform [Sebe12] as shown in Figure 14-1.

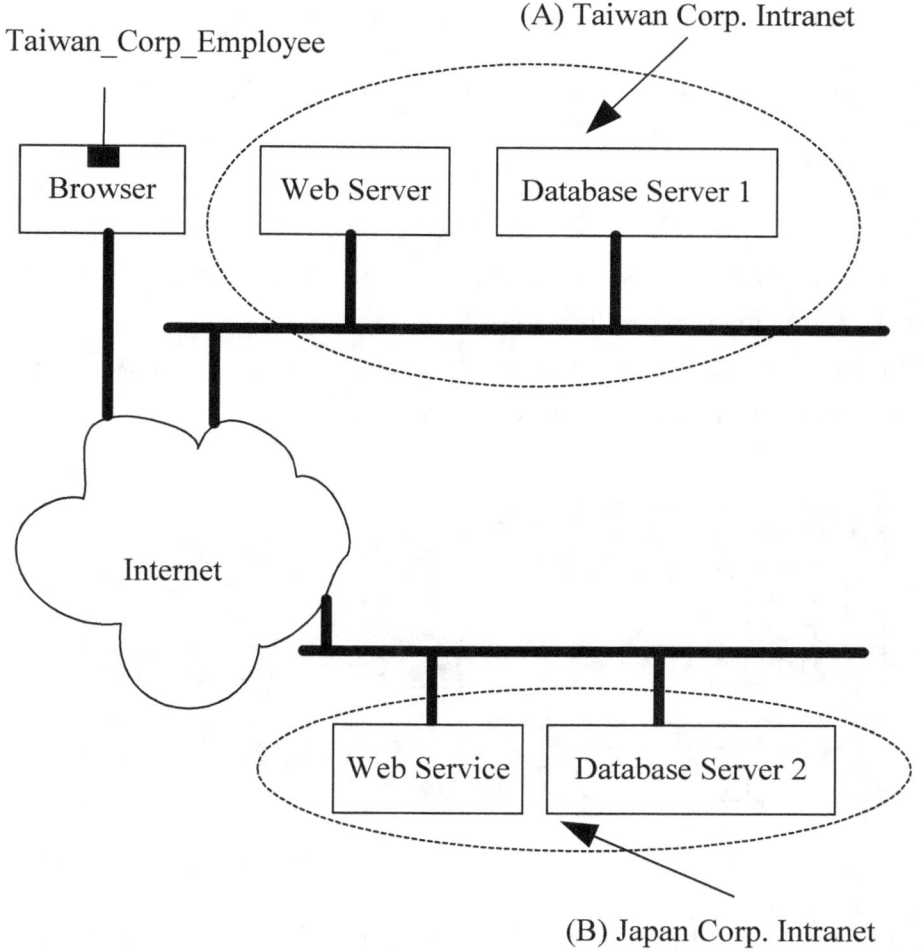

Figure 14-1　*Web Service Extranet System* on a Web Multi-Tier Platform

The major functionality of the *Web Service Extranet System* is to provide a web browser for the *Taiwan_Corp_Employee* actor to input the *PurchaseDate* value as shown in Figure 14-2.

132

Input the Purchase Date

Figure 14-2   Web Browser for the Actor to input the *Purchase Date*

After the *Taiwan_Corp_Employee* actor inputs the *PurchaseDate* value and presses down the *ExtranetButton* button then the *Web Service Extranet System* generates the following two results:

(A) The first result is to insert a new record into the *PurchaseTable* table of the *Taiwan_Database* database as shown in Figure 14-3.

Inserting a new record

Figure 14-3   First Result After Pressing Down the *ExtranetButton* Button

(B) The second result is to insert a new record into the *SaleTable* table of the *Japan_Database* database as shown in Figure 14-4.

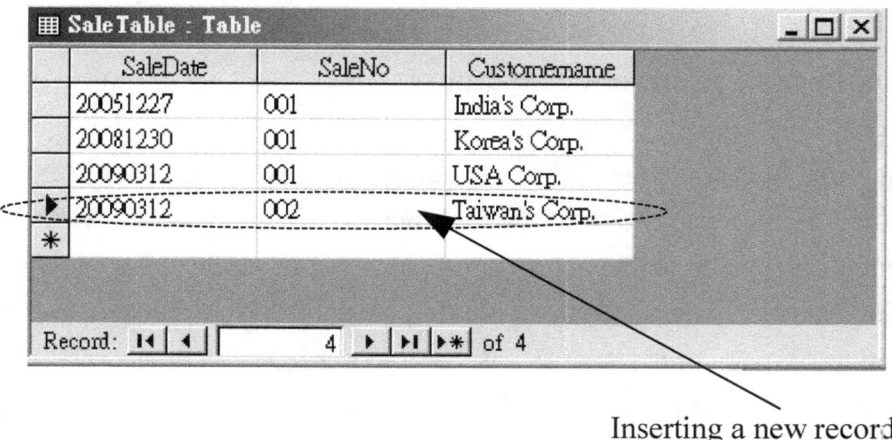

Inserting a new record

Figure 14-4   Second Result After Pressing Down the *ExtranetButton* Button

Finally, we also see a message informing us that the Extranet deal is done, on the web page, as shown in Figure 14-5.

ExtraNet is Done

Figure 14-5   Informing that the Extranet Deal is Done

Using the SBC architecture description language, we shall go through: a) architecture hierarchy diagram, b) framework diagram, c) component operation diagram, d) component connection diagram, e) structure-behavior coalescence diagram and f) interaction flow diagram, to accomplish systems thinking 2.0 (architectural thinking) defining the *web service extranet system*.

14-1 Architecture Hierarchy Diagram

We use an architecture hierarchy diagram (AHD) to define the multi-level composition and decomposition of the *web service extranet system*. AHD is the first fundamental diagram to achieve structure-behavior coalescence. As shown in Figure 14-6, *Web Service Extranet System* is composed of *WebServerExtraNet_GUI* and *W_Subsystem_2*; *W_Subsystem_2* is composed of *Taiwan_Database* and *W_Subsystem_1*; *W_Subsystem_1* is composed of *SaleComplete_Logic* and *Japan_Database*.

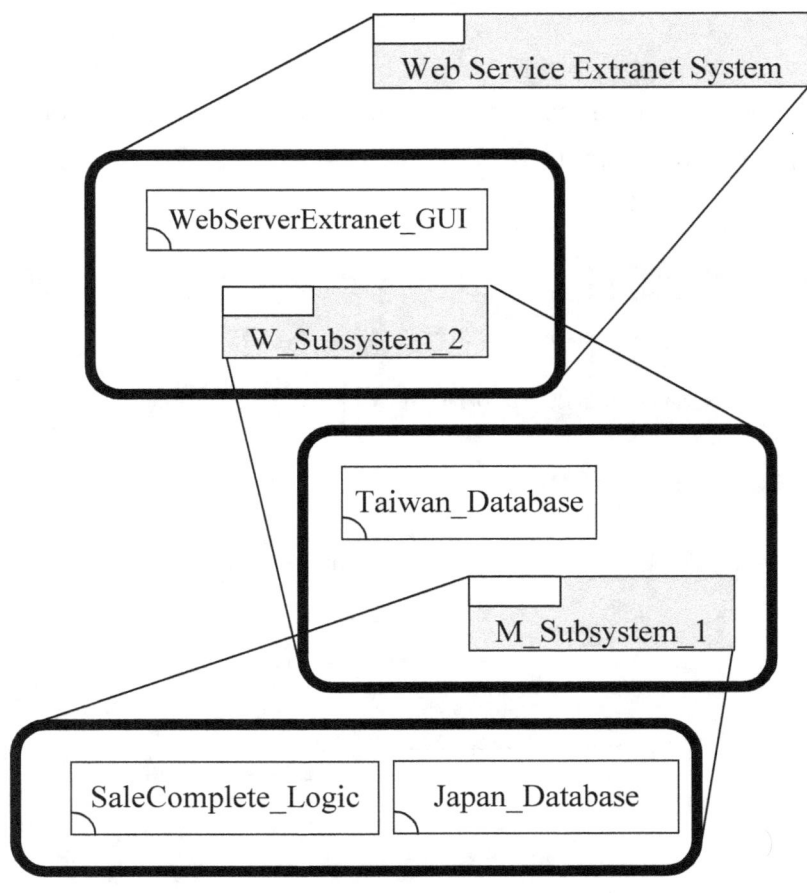

Figure 14-6   AHD of the *Web Service Extranet System*

In Figure 14-6, *Web Service Extranet System*, *W_Subsystem_2* and *W_Subsystem_1* are aggregated systems while *WebServerExtranet_GUI*,

Taiwan_Database, *SaleComplete_Logic* and *Japan_Database* are non-aggregated systems.

14-2 Framework Diagram

We use a framework diagram (FD) to define the multi-layer composition and decomposition of the *web service extranet system* as shown in Figure 14-7. FD is the second fundamental diagram to achieve structure-behavior coalescence. In the figure, *Application_Layer* contains the *WebServerExtranet_GUI* component; *Logic_Layer* contains the *SaleComplete_Logic* component; *Data_Layer* contains the *Taiwan_Database* and *Japan_Database* components.

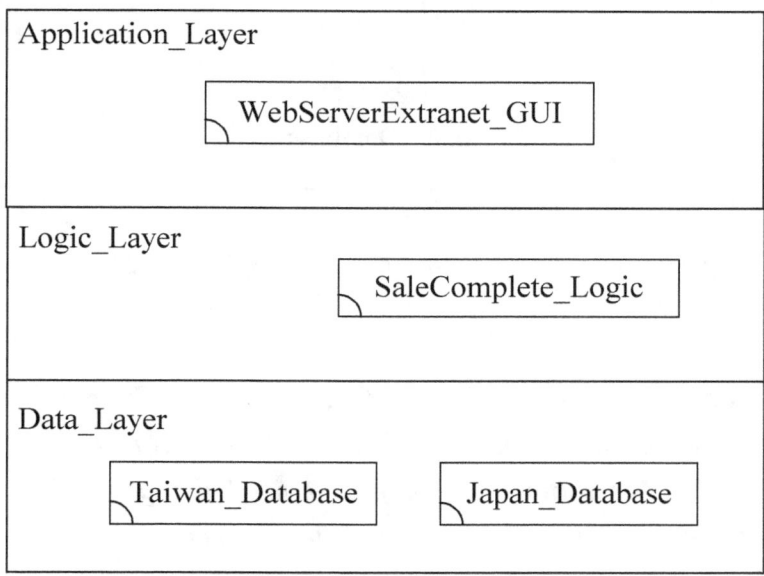

Figure 14-7   FD of the *Web Service Extranet System*

14-3 Component Operation Diagram

We use a component operation diagram (COD) to define the operations of all components of the *web service extranet system* as shown in Figure 14-8. COD is the third fundamental diagram to achieve structure-behavior coalescence. In the figure, component *WebServerExtranet_GUI* has one operation: *ExtranetButton_Click*; component *SaleComplete_Logic* has one operation: *Add_a_Sale*; component *Taiwan_Database* has one operation: *Sql_p_Insert*; component *Japan_Database* has one operation: *Sql_s_Insert*.

136

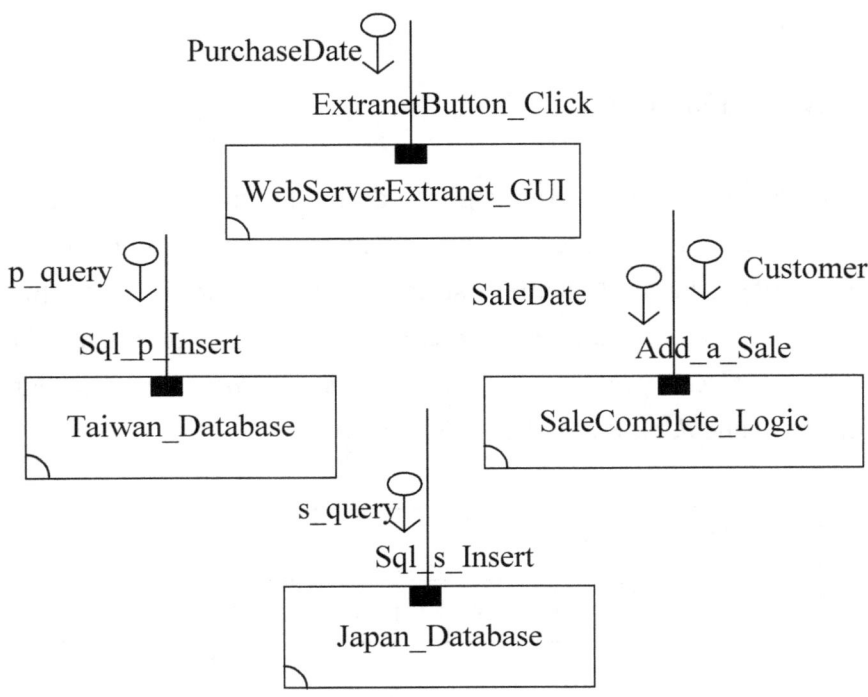

Figure 14-8　COD of the *Web Service Extranet System*

The operation formula of *ExtranetButton_Click* is *ExtranetButton_Click(In PurchaseDate)*. The operation formula of *Add_a_Sale* is *Add_a_Sale(In SaleDate, Customer)*. The operation formula of *Sql_p_Insert* is *Sql_p_Insert(In p_query)*. The operation formula of *Sql_s_Insert* is *Sql_s_Insert(In s_query)*.

Figure 14-9 shows the primitive data type specification of the *PurchaseDate*, *SaleDate* and *Customer* input parameters.

Parameter	Data Type	Instances
PurchaseDate	String	20090112
SaleDate	String	20090112
Customer	String	Taiwan Corp.

Figure 14-9　Primitive Data Type Specification

Figure 14-10 shows the composite data type specification of the *p_query* input parameter occurring in the *Sql_p_Insert(In p_query)* operation formula.

Parameter	p_query

Data Type	TABLE of
	  Purchase Date : Text
	  Purchase No : Text
	  Supplier : Text
	End TABLE ;

Instances	<table><tr><td>PurchaseDate</td><td>PurchaseNo</td><td>Supplier</td></tr><tr><td>20090312</td><td>001</td><td>Japan Corp.</td></tr></table>

Figure 14-10   Composite Data Type Specification

Figure 14-11 shows the composite data type specification of the *s_query* input parameter occurring in the *Sql_s_Insert(In s_query)* operation formula.

Parameter	s_query

Data Type	TABLE of
	  SaleDate : Text
	  SaleNo : Text
	  Customer : Text
	End TABLE ;

Instances	<table><tr><td>Sale Date</td><td>Sale No</td><td>Customer</td></tr><tr><td>20090312</td><td>001</td><td>Taiwan Corp.</td></tr></table>

Figure 14-11   Composite Data Type Specification

14-4 Component Connection Diagram

We use a component connection diagram (CCD) to define the connections among components and actors of the *web service extranet system* as shown in Figure

138

14-12. CCD is the fourth fundamental diagram to achieve structure-behavior coalescence.

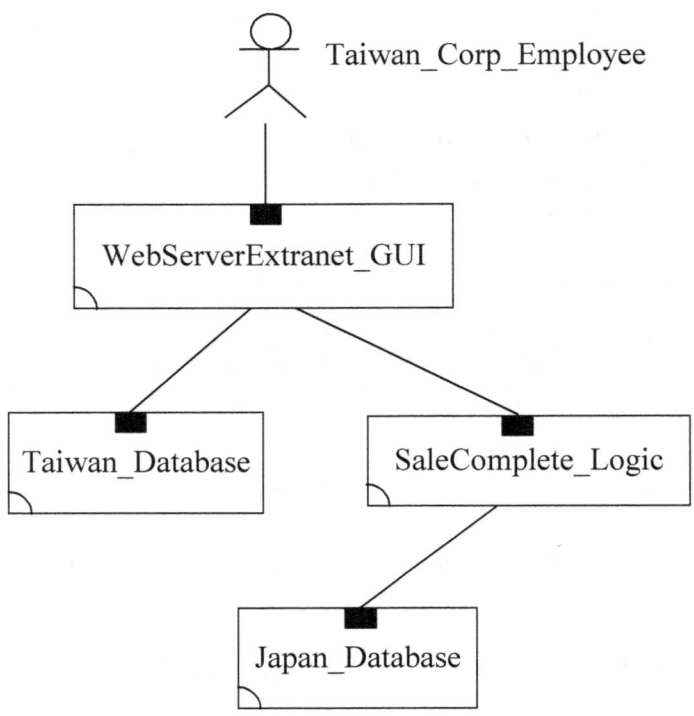

Figure 14-12   CCD of the *Web Service Extranet System*

In Figure 14-12, actor *Taiwan_Corp_Employee* has one connection with the *WebServerExtranet_GUI* component; component *WebServerExtranet_GUI* has one connection with each of the *Taiwan_Database* and *SaleComplete_Logic* components; component *SaleComplete_Logic* has one connection with the *Japan_Database* component.

14-5 Structure-Behavior Coalescence Diagram

We use a structure-behavior coalescence diagram (SBCD) to define the systems structure and systems behavior coexisting in the *web service extranet system* as shown in Figure 14-13. SBCD is the fifth fundamental diagram to achieve structure-behavior coalescence.

In the figure, interactions among the *Taiwan_Corp_Employee* actor and the *WebServerExtranet_GUI*, *Taiwan_Database*, *SaleComplete_Logic*, *Japan_Database* components shall draw forth the *Purchase&Sale* behavior.

Figure 14-13   SBCD of the *Web Service Extranet System*

The overall behavior of a system is the aggregation of all its individual behaviors. For example, the overall behavior of the *web service extranet system* includes the *Purchase&Sale* behavior.

The major purpose of SBC architecture description language is to define an integrated whole of a system. In Figure 14-13, we not only see its structure, but also see at the same time its behavior in the *web service extranet system's* SBCD.

14-6 Interaction Flow Diagram

The overall behavior of the *web service extranet system* includes only one individual behavior: *Purchase&Sale*. This individual behavior is represented by an execution path. We use an interaction flow diagram (IFD) to define this execution path. IFD is the sixth fundamental diagram to achieve structure-behavior coalescence.

Figure 14-14 shows an IFD of the *Purchase&Sale* behavior. First, actor *Taiwan_Corp_Employee* interacts with the *WebServernExtranet_GUI* component through the *ExtranetButton_Click* operation call interaction, carrying the

PurchaseDate input parameter. Next, component *WebServerExtranet_GUI* interacts with the *Taiwan_Database* component through the *Sql_p_insert* operation call interaction, carrying the *p_query* input parameter. Continuingly, component *WebServerExtranet_GUI* interacts with the *SaleComplete_Logic* component through the *Add_A_Sale* operation call interaction, carrying the *SaleDate* and *CustomerName* input parameters. Continuingly, component *SaleComplete_Logic* interacts with the *Japan_Database* component through the *Sql_s_insert* operation call interaction, carrying the *s_query* input parameter. Finally, actor *Taiwan_Corp_Employee* interacts with the *WebServerArithmetic_GUI* component through the *ExtranetButton_Click* operation return interaction.

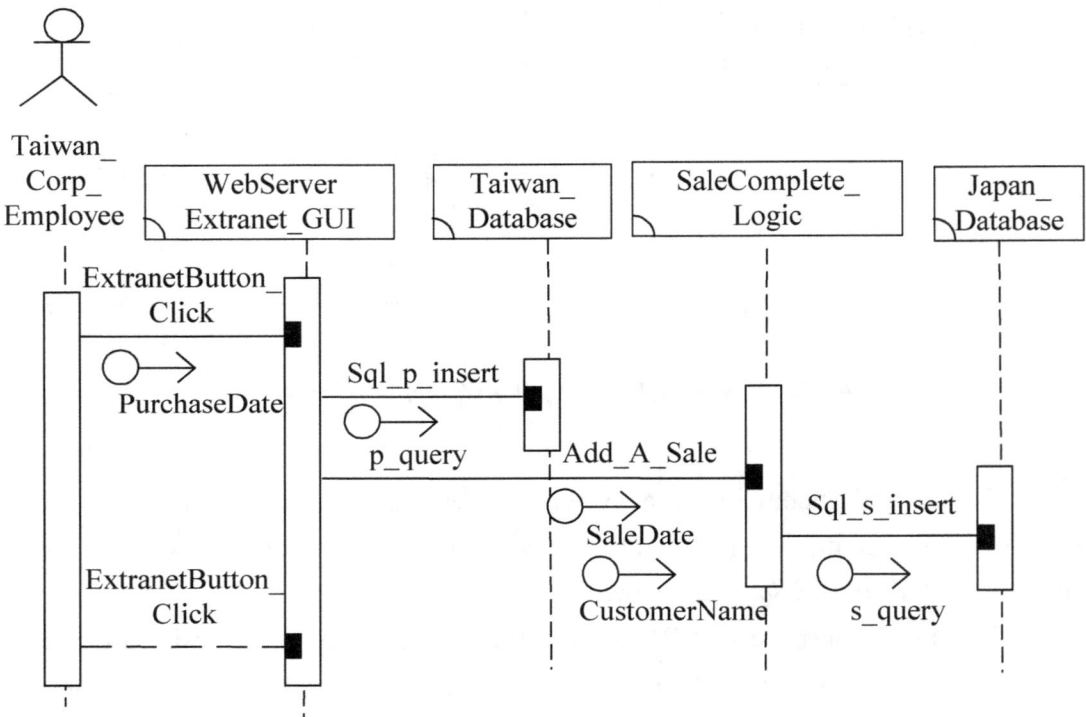

Figure 14-14   IFD of the *Purchase&Sale* Behavior

Chapter 15: Systems Thinking 2.0 Defining the Convenience Store's Get 2nd 50% off Sales Promotion System

This chapter demonstrates how to achieve systems thinking 2.0 (architectural thinking) defining the *convenience store's get 2nd 50% off sales promotion system*, through the application of SBC architecture description language (SBC-ADL).

In order to attract customers to buy, a convenience store may find many ways to do so. The easiest one is to give discounts. Because of human nature, coupled with the hot summer weather, the customers see the advertisement saying a cold drink at 50% off, they may have a great desire to buy it. This inspires the idea of a *convenience store's get 2nd 50% off sales promotion system*.

The overall behavior of the convenience store's getting 2nd 50% off sales promotion system is represented by two individual behaviors: *Advertising* and *Get_2nd_50%_off* as shown in Figure 15-1.

Figure 15-1   Two Behaviors of
the *Convenience Store's Get 2nd 50% off Sales Promotion System*

Using the SBC architecture description language, we shall go through: a) architecture hierarchy diagram, b) framework diagram, c) component operation diagram, d) component connection diagram, e) structure-behavior coalescence diagram and f) interaction flow diagram, to accomplish systems thinking 2.0 (architectural thinking) defining the *convenience store's get 2nd 50% off sales promotion system*.

142

15-1 Architecture Hierarchy Diagram

We use an architecture hierarchy diagram (AHD) to define the multi-level composition and decomposition of the *convenience store's get 2nd 50% off sales promotion system*. AHD is the first fundamental diagram to achieve structure-behavior coalescence. As shown in Figure 15-2, *Convenience Store's Get 2nd 50% off Sales Promotion System* is composed of *Clerk* and *Subsystem_1*; *Subsystem_1* is composed of *Advertising_Board*.

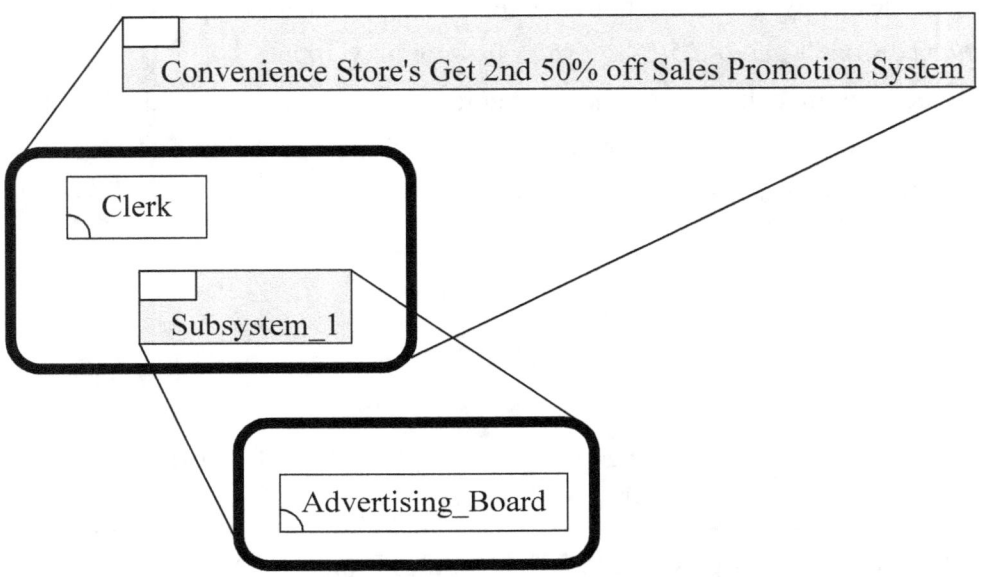

Figure 15-2   AHD of the *Convenience Store's Get 2nd 50% off Sales Promotion System*

In Figure 15-2, *Convenience Store's Get 2nd 50% off Sales Promotion System* and *Subsystem_1* are aggregated systems while *Clerk* and *Advertising_Board* are non-aggregated systems.

15-2 Framework Diagram

We use a framework diagram (FD) to define the multi-layer composition and decomposition of the *convenience store's get 2nd 50% off sales promotion system* as shown in Figure 15-3. FD is the second fundamental diagram to achieve structure-behavior coalescence. In the figure, *Business Layer* contains the *Clerk* component;

Application Layer contains the *Advertising_Board* component.

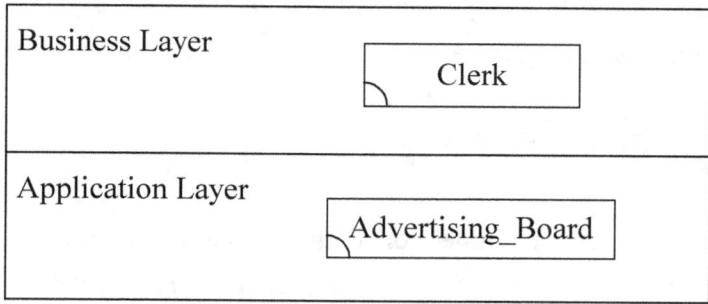

Figure 15-3   FD of the *Convenience Store's Get 2nd 50% off Sales Promotion System*

15-3 Component Operation Diagram

We use a component operation diagram (COD) to define the operations of all components of the *convenience store's get 2nd 50% off sales promotion system* as shown in Figure 15-4. COD is the third fundamental diagram to achieve structure-behavior coalescence. In the figure, component *Advertising_Board* has one operation: *Watch*; component *Clerk* has two operations: *Buy_the_1st_Goods* and *Buy_the_2nd_Goods*.

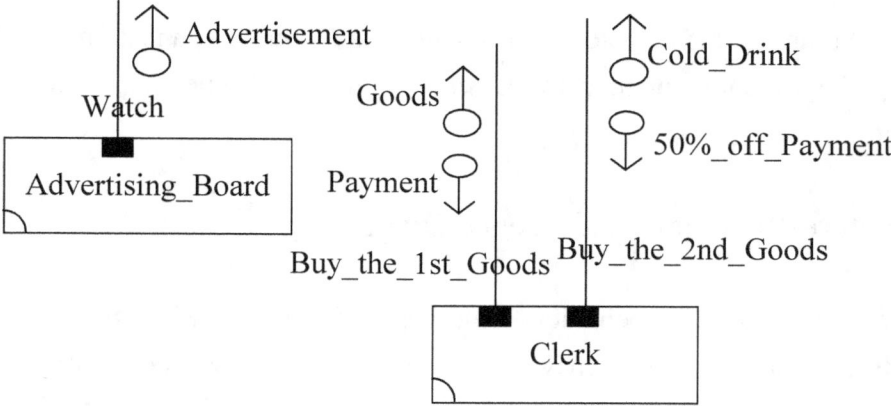

Figure 15-4   COD of the *Convenience Store's Get 2nd 50% off Sales Promotion System*

The operation formula of *Watch* is *Watch(Out Advertisement)*. The operation formula of *Buy_the_1st_Goods* is *Buy_the_1st_Goods(In Payment; Out Goods)*. The operation formula of *Buy_the_2nd_Goods* is *Buy_the_2nd_Goods(In 50%_off_Payment; Out Cold_Drink)*.

15-4 Component Connection Diagram

We use a component connection diagram (CCD) to define the connections among components and actors of the *convenience store's get 2nd 50% off sales promotion system* as shown in Figure 15-5. CCD is the fourth fundamental diagram to achieve structure-behavior coalescence.

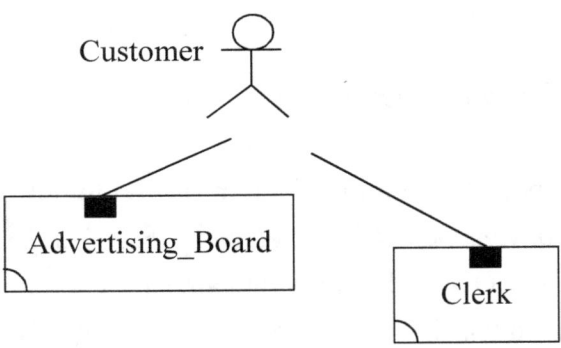

Figure 15-5   CCD of the *Convenience Store's Get 2nd 50% off Sales Promotion System*

In Figure 15-5, actor *Customer* has one connection with the *Advertising_Board* component; actor *Customer* has one connection with the *Clerk* component.

15-5 Structure-Behavior Coalescence Diagram

We use a structure-behavior coalescence diagram (SBCD) to define the systems structure and systems behavior coexisting in the *convenience store's get 2nd 50% off sales promotion system* as shown in Figure 15-6. SBCD is the fifth fundamental diagram to achieve structure-behavior coalescence. In the figure, interactions among the *Customer* actor and the *Advertising_Board, Clerk* components shall draw forth the *Advertising* and *Get_2nd_50%_off* behaviors.

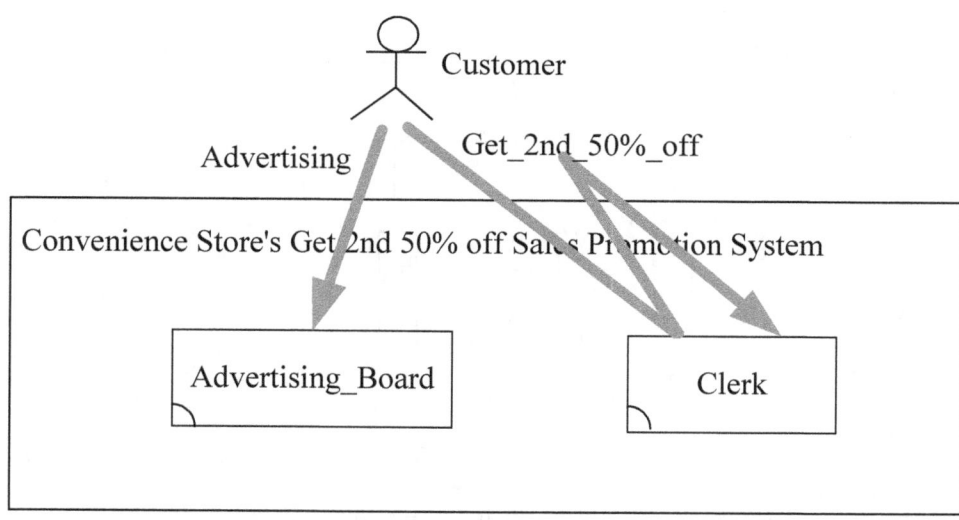

Figure 15-6   SBCD of the *Convenience Store's Get 2nd 50% off Sales Promotion System*

The overall behavior of a system is the aggregation of all its individual behaviors. For example, the overall behavior of the *convenience store's get 2nd 50% off sales promotion system* includes the *Advertising* and *Get_2nd_50%_off* behavior.

Be noticed that the *Advertising* and *Get_2nd_50%_off* behaviors are mutually independent of each other. They tend to be executed concurrently [Hoar85, Miln89, Miln99].

The major purpose of SBC architecture description language is to define an integrated whole of a system. In Figure 15-6, we not only see its structure, but also see at the same time its behavior in the *convenience store's get 2nd 50% off sales promotion system*'s SBCD.

15-6 Interaction Flow Diagram

The overall behavior of the *convenience store's get 2nd 50% off sales promotion system* includes two individual behaviors: *Advertising* and *Get_2nd_50%_off*. Each individual behavior is represented by an execution path. We use an interaction flow diagram (IFD) to define each one of these execution paths. IFD is the sixth fundamental diagram to achieve structure-behavior coalescence.

Figure 15-7 shows an IFD of the *Advertising* behavior. First, actor *Customer* interacts with the *Advertising_Board* component through the *Watch* operation call interaction, carrying the *Advertisement* output parameter.

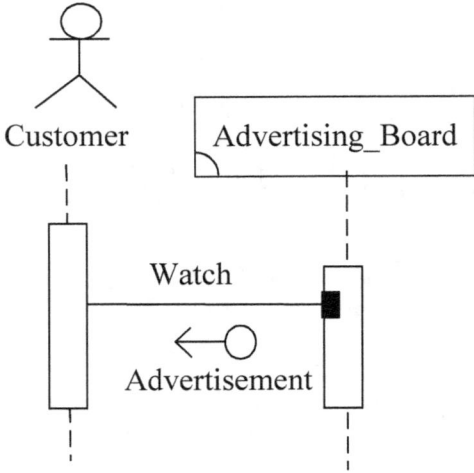

Figure 15-7   IFD of the *Advertising* Behavior

Figure 15-8 shows an IFD of the *Get_2nd_50%_off* behavior. First, actor *Customer* interacts with the *Clerk* component through the *Buy_the_1st_Goods* operation call interaction, carrying the *Payment* input parameter and *Goods* output parameter. Last, actor *Customer* interacts with the *Clerk* component through the *Buy_the_2nd_Goods* operation call interaction, carrying the *50%_off_Payment* input parameter and *Cold_Drink* output parameter.

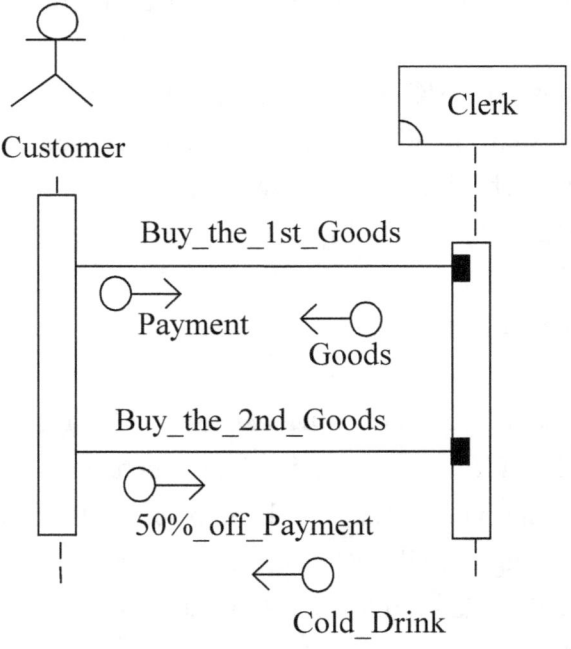

Figure 15-8   IFD of the *Get_2nd_50%_off* Behavior

Chapter 16: Systems Thinking 2.0 Defining the Department Store's Car Sweepstakes Sales Promotion System

This chapter demonstrates how to achieve systems thinking 2.0 (architectural thinking) defining the *department store's car sweepstakes sales promotion system*, through the application of SBC architecture description language (SBC-ADL).

A decent car is something everyone wants to have. A customer will be interested in buying a thousand dollars goods in a department store if he has the opportunity to get a grand new one-million-dollars car. This inspires the idea of a *department store's car sweepstakes sales promotion system*.

The overall behavior of the *department store's car sweepstakes sales promotion system* is represented by three individual behaviors: *Get_Sweepstake_Number*, *Draw_Out_the_Winners_List* and *Collect_the_Winning_Car* as shown in Figure 16-1.

Figure 16-1   Three Behaviors of
the *Department Store's Car Sweepstakes Sales Promotion System*

Using the SBC architecture description language, we shall go through: a) architecture hierarchy diagram, b) framework diagram, c) component operation diagram, d) component connection diagram, e) structure-behavior coalescence diagram and f) interaction flow diagram, to accomplish systems thinking 2.0 (architectural thinking) defining the *department store's car sweepstakes sales promotion system*.

16-1 Architecture Hierarchy Diagram

We use an architecture hierarchy diagram (AHD) to define the multi-level composition and decomposition of the *department store's car sweepstakes sales promotion system*. AHD is the first fundamental diagram to achieve structure-behavior coalescence. As shown in Figure 16-2, *Department Store's Car Sweepstakes Sales Promotion System* is composed of *Clerk, Sweepstake_Host, General_Manager* and *Subsystem_1*; *Subsystem_1* is composed of *Sweepstake_GUI*.

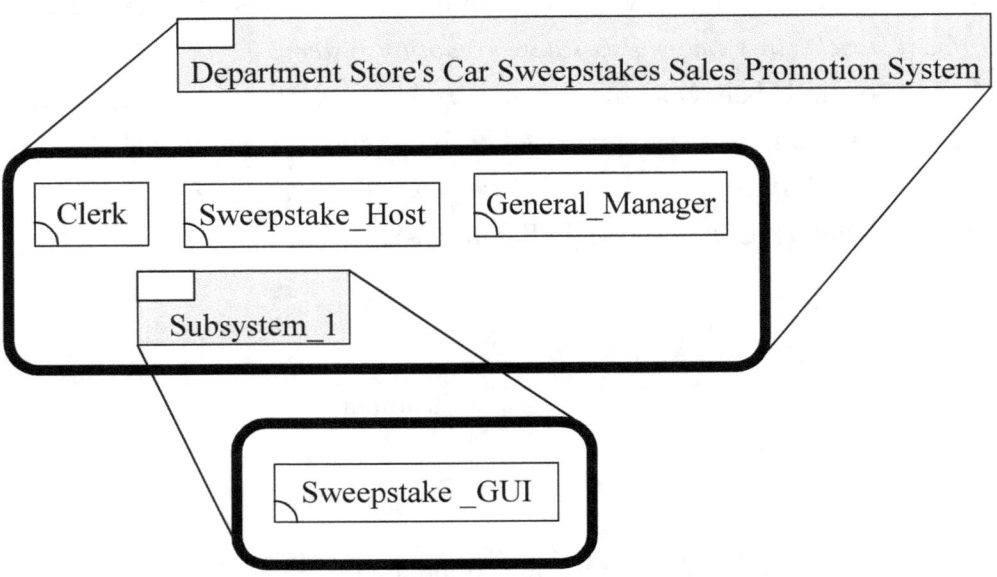

Figure 16-2   AHD of the *Department Store's Car Sweepstakes Sales Promotion System*

In Figure 16-2, *Department Store's Car Sweepstakes Sales Promotion System* and *Subsystem_1* are aggregated systems while *Clerk, Sweepstake_Host, General_Manager* and *Sweepstake_GUI* are non-aggregated systems.

16-2 Framework Diagram

We use a framework diagram (FD) to define the multi-layer composition and decomposition of the *department store's car sweepstakes sales promotion system* as shown in Figure 16-3. FD is the second fundamental diagram to achieve structure-behavior coalescence. In the figure, *Business_Layer* contains the *Clerk, Sweepstake_Host* and *General_Manager* components; *Application_Layer* contains the *Sweepstake_GUI* component.

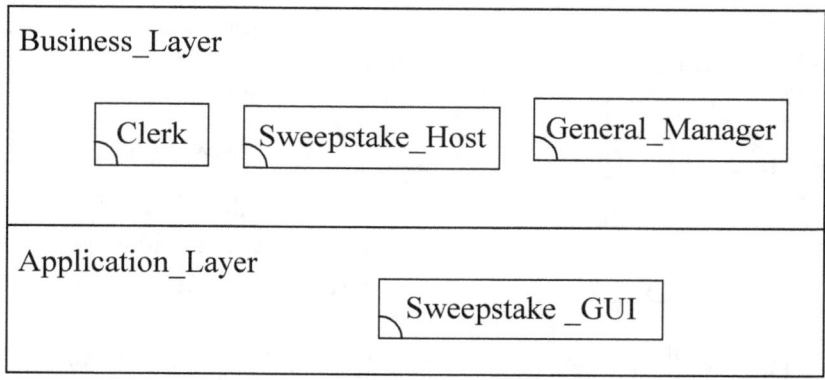

Figure 16-3   FD of the *Department Store's Car Sweepstakes
Sales Promotion System*

16-3 Component Operation Diagram

We use a component operation diagram (COD) to define the operations of all components of the *department store's car sweepstakes sales promotion system* as shown in Figure 16-4. COD is the third fundamental diagram to achieve structure-behavior coalescence. In the figure, component *Clerk* has one operation: *Buy_More_Than_Thousand_Dollars*; component *Sweepstake_Host* has one operation: Draw_Out; component *General_Manager* has one operation: *Award*; component *Sweepstake_GUI* has two operations: *Register* and *Lucky_Draw*.

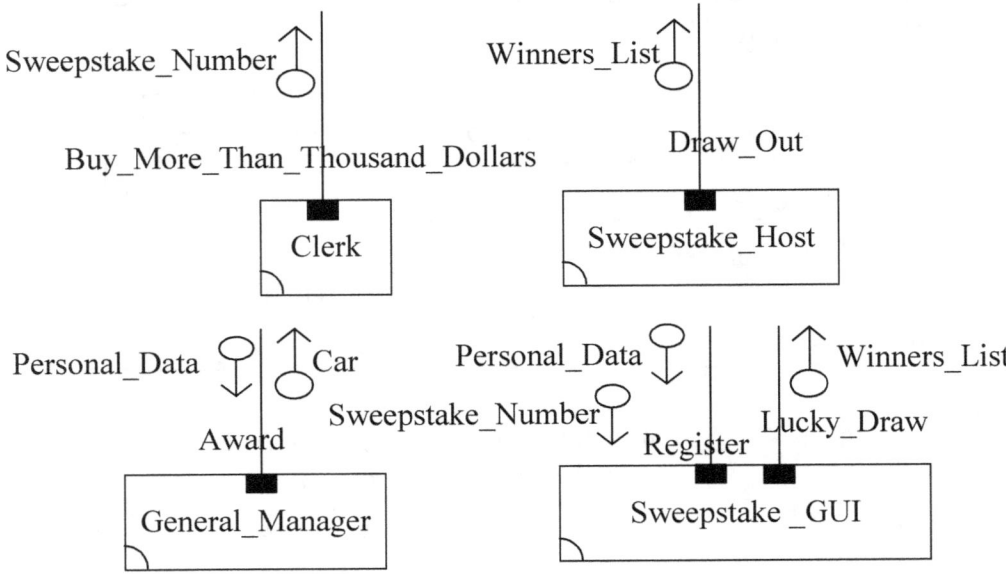

Figure 16-4   COD of the *Department Store's Car Sweepstakes Sales Promotion System*

The operation formula of *Buy_More_Than_Thousand_Dollars* is *Buy_More_Than_Thousand_Dollars(Out Sweepstake_Number)*. The operation formula of Draw_Out is Draw_Out*(Out Winners_List)*. The operation formula of *Award* is *Award(In Personal_Data; Out Car)*. The operation formula of *Register* is *Register(In Sweepstake_Number, Personal_Data)*. The operation formula of *Lucky_Draw* is *Lucky_Draw(Out Winners_List)*.

16-4 Component Connection Diagram

We use a component connection diagram (CCD) to define the connections among components and actors of the *department store's car sweepstakes sales promotion system* as shown in Figure 16-5. CCD is the fourth fundamental diagram to achieve structure-behavior coalescence.

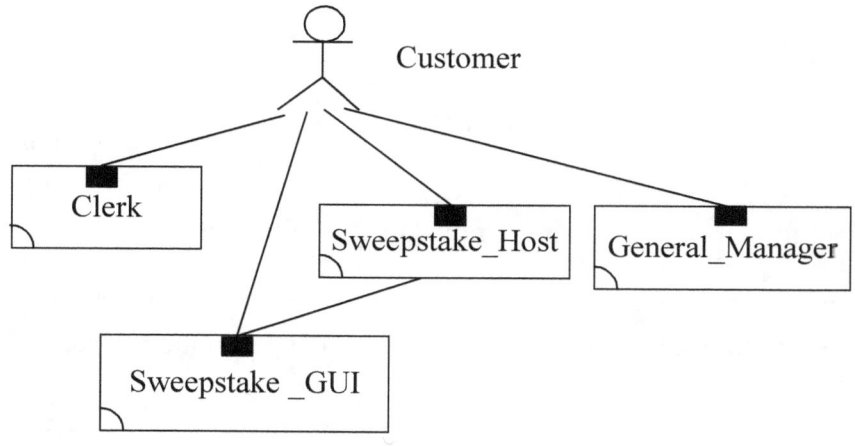

Figure 16-5   CCD of the *Department Store's Car Sweepstakes Sales
Promotion System*

In Figure 16-5, actor *Customer* has one connection with each of the *Clerk,
Sweepstake_Host*, *General_Manager* and *Sweepstake_GUI* components; component
Sweepstake_Host has one connection with the *Sweepstake_GUI* component.

16-5 Structure-Behavior Coalescence Diagram

We use a structure-behavior coalescence diagram (SBCD) to define the
systems structure and systems behavior coexisting in the *department store's car
sweepstakes sales promotion system* as shown in Figure 16-6. SBCD is the fifth
fundamental diagram to achieve structure-behavior coalescence. In the figure,
interactions among the *Customer* actor and the *Clerk, Sweepstake_GUI,
Sweepstake_Host, General_Manager* components shall draw forth the
Get_Sweepstake_Number,   *Draw_Out_the_Winners_List*   and
Collect_the_Winning_Car behaviors.

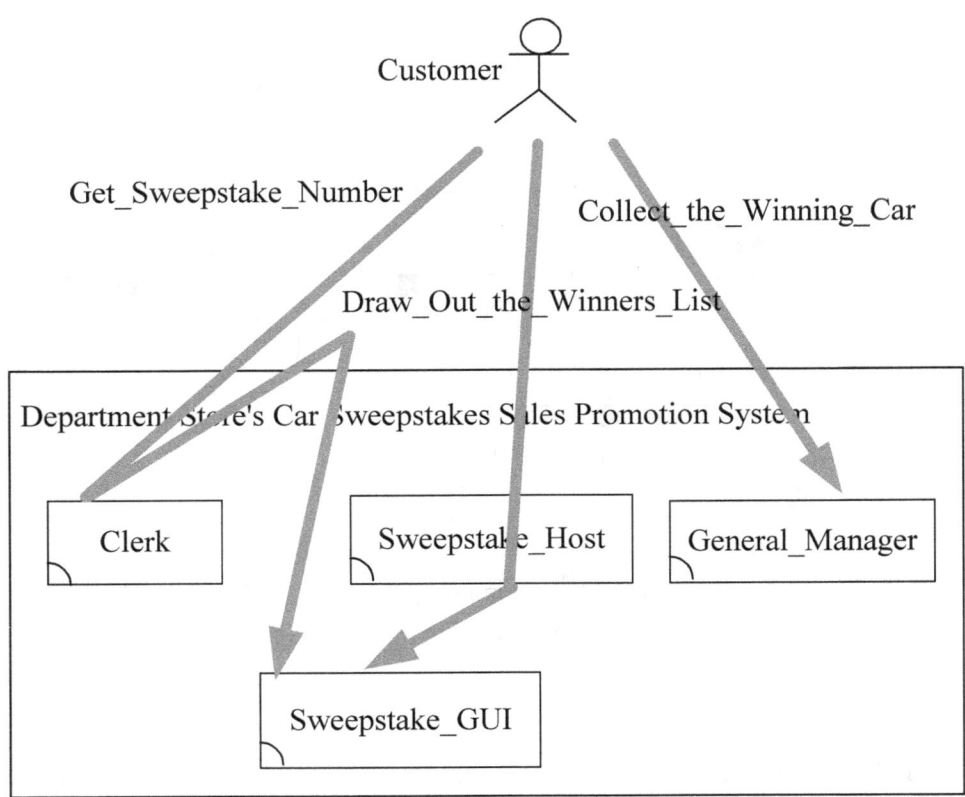

Figure 16-6   SBCD of the *Department Store's Car Sweepstakes Sales Promotion System*

The overall behavior of a system is the aggregation of all its individual behaviors. For example, the overall behavior of the *department store's car sweepstakes sales promotion system* includes the *Get_Sweepstake_Number*, *Draw_Out_the_Winners_List* and *Collect_the_Winning_Car* behaviors.

Be noticed that the *Get_Sweepstake_Number*, *Draw_Out_the_Winners_List* and *Collect_the_Winning_Car* behaviors are mutually independent of each other. They tend to be executed concurrently [Hoar85, Miln89, Miln99].

The major purpose of SBC architecture description language is to define an integrated whole of a system. In Figure 16-6, we not only see its structure, but also see at the same time its behavior in the *department store's car sweepstakes sales promotion system*'s SBCD

16-6 Interaction Flow Diagram

The overall behavior of the *department store's car sweepstakes sales promotion system* includes three individual behaviors: *Get_Sweepstake_Number*,

Draw_Out_the_Winners_List and *Collect_the_Winning_Car*. Each individual behavior is represented by an execution path. We use an interaction flow diagram (IFD) to define each one of these execution paths. IFD is the sixth fundamental diagram to achieve structure-behavior coalescence.

Figure 16-7 shows an IFD of the *Get_Sweepstake_Number* behavior. First, actor *Customer* interacts with the *Clerk* component through the *Buy_More_Than_Thousand_Dollars* operation call interaction, carrying the *Sweepstake_Number* output parameter. Last, actor *Customer* interacts with the *Sweepstake_GUI* component through the *Register* operation call interaction, carrying the *Personal_Data* and *Sweepstake_Number* input parameters.

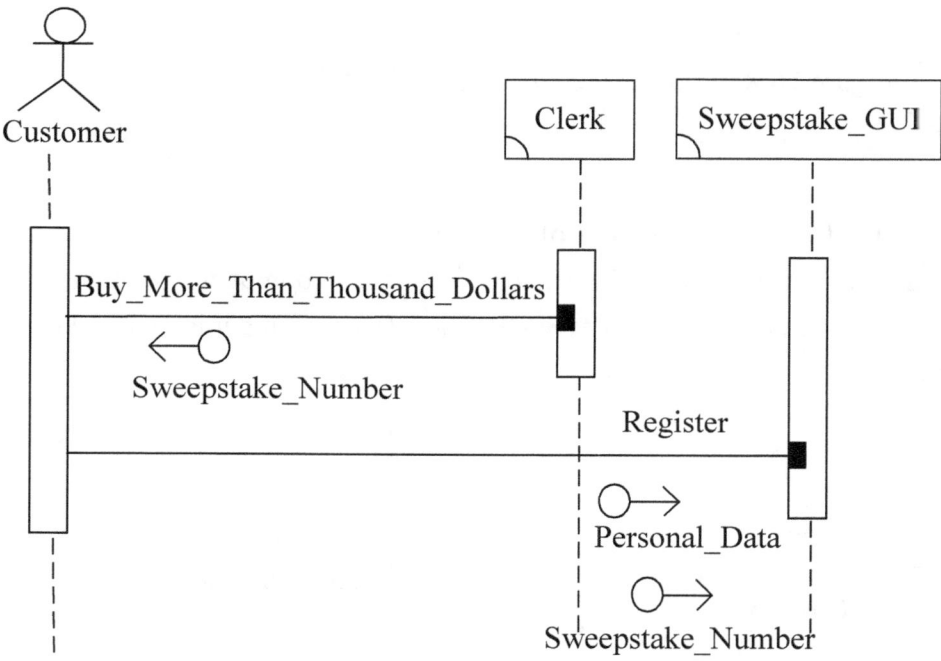

Figure 16-7   IFD of the *Get_Sweepstake_Number* Behavior

Figure 16-8 shows an IFD of the *Draw_Out_the_Winners_List* behavior. First, actor *Customer* interacts with the *Sweepstake_Ho*st component through the *Draw_Out* operation call interaction.   Next, component *Sweepstake_Host* interacts with the *Sweepstake_GUI* component through the *Lucky_Draw* operation call interaction, carrying the *Winners_List* output parameter. Last, actor *Customer* interacts with the *Sweepstake_Host* component through the *Draw_Out* operation return interaction, carrying the *Winners_List* output parameter.

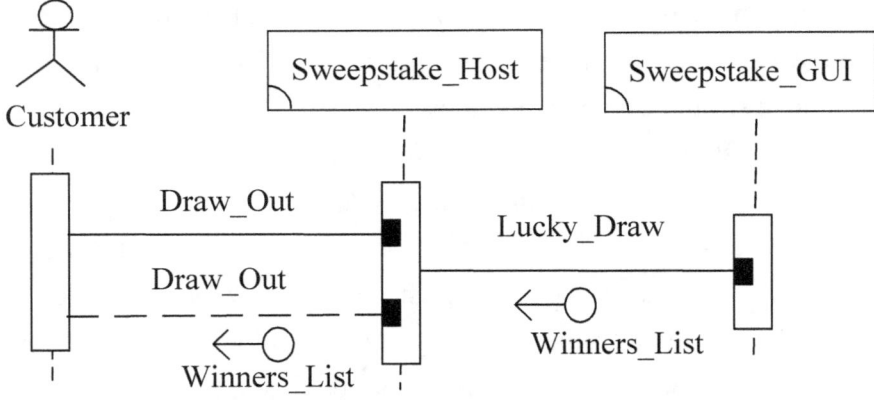

Figure 16-8   IFD of the *Draw_Out_the_Winners_List* Behavior

Figure 16-9 shows an IFD of the *Collect_the_Winning_Car* behavior. First, actor *Customer* interacts with the *General_Manager* component through the *Award* operation call interaction, carrying the *Personal_Data* input parameter and *Car* output parameter.

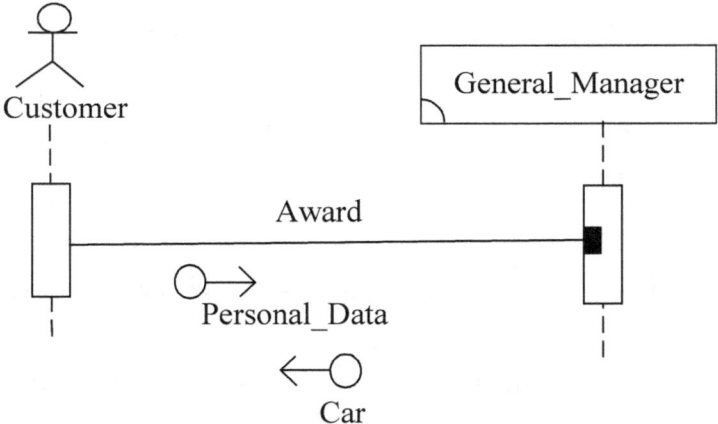

Figure 16-9   IFD of the *Collect_the_Winning_Car* Behavior

Chapter 17: Systems Thinking 2.0 Defining the Female Mantis

This chapter demonstrates how to achieve systems thinking 2.0 (architectural thinking) defining the *female mantis*, through the application of SBC architecture description language (SBC-ADL).

In general, the overall behavior of the *female mantis* is prominently represented by three individual behaviors: *Preying*, *Walking* and *Reproducing* as shown in Figure 17-1.

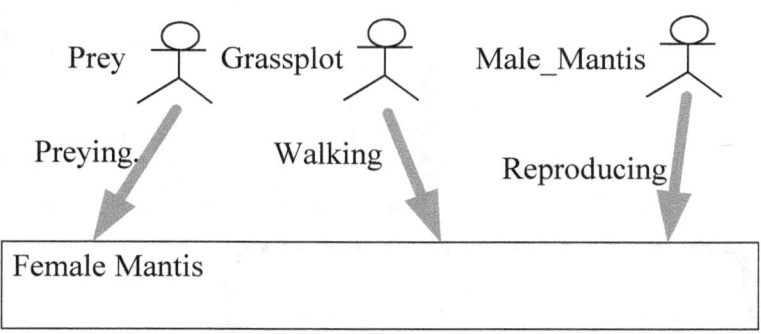

Figure 17-1   Three Behaviors of the *Female Mantis*

Using the SBC architecture description language, we shall go through: a) architecture hierarchy diagram, b) framework diagram, c) component operation diagram, d) component connection diagram, e) structure-behavior coalescence diagram and f) interaction flow diagram, to accomplish systems thinking 2.0 (architectural thinking) defining the *female mantis*.

17-1 Architecture Hierarchy Diagram

We use an architecture hierarchy diagram (AHD) to define the multi-level composition and decomposition of the *female mantis*. AHD is the first fundamental diagram to achieve structure-behavior coalescence. As shown in Figure 17-2, *Female Mantis* is composed of *Forelegs*, *Genital_Opening* and *Subsystem_2*; *Subsystem_2* is composed of *Mouthparts*, *Hind_Legs* and *Subsystem_1*; *Subsystem_1* is composed of *Digestive_System* and *Eggs*.

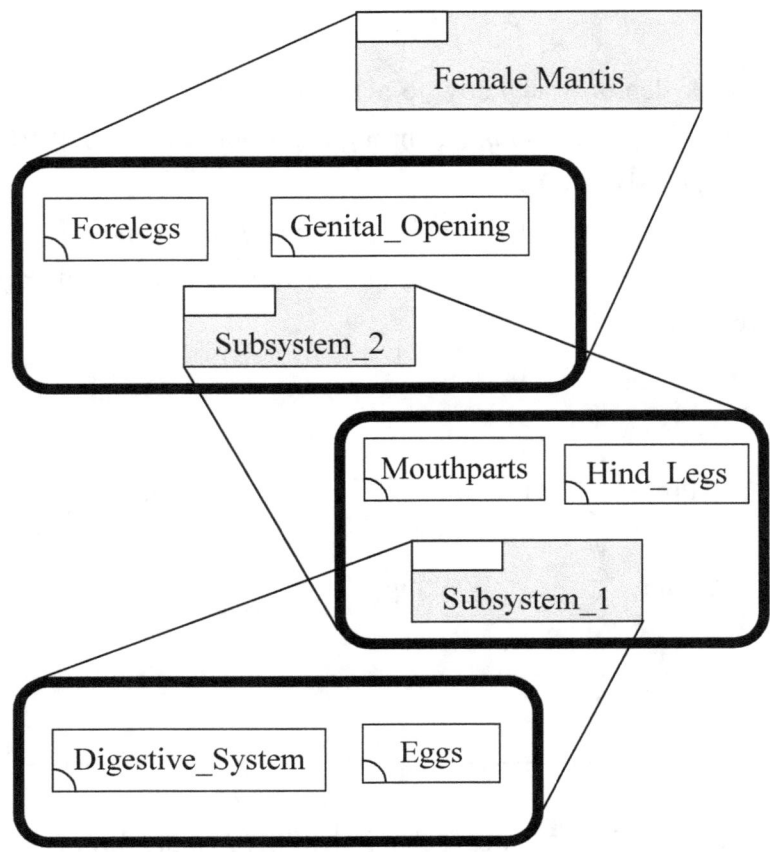

Figure 17-2   AHD of the *Female Mantis*

In Figure 17-2, *Female Mantis*, *Subsystem_2* and *Subsystem_1* are aggregated systems while *Forelegs*, *Genital_Opening*, *Mouthparts*, *Hind_Legs*, *Digestive_System* and *Eggs* are non-aggregated systems.

17-2 Framework Diagram

We use a framework diagram (FD) to define the multi-layer composition and decomposition of the *female mantis* as shown in Figure 17-3. FD is the second fundamental diagram to achieve structure-behavior coalescence. In the figure, *Technology_SubLayer_3* contains the *Forelegs* and *Genital_Opening* components; *Technology_SubLayer_2* contains the *Mouthparts* and *Hind_Legs* components; *Technology_SubLayer_1* contains the *Digestive_System* and *Eggs* components.

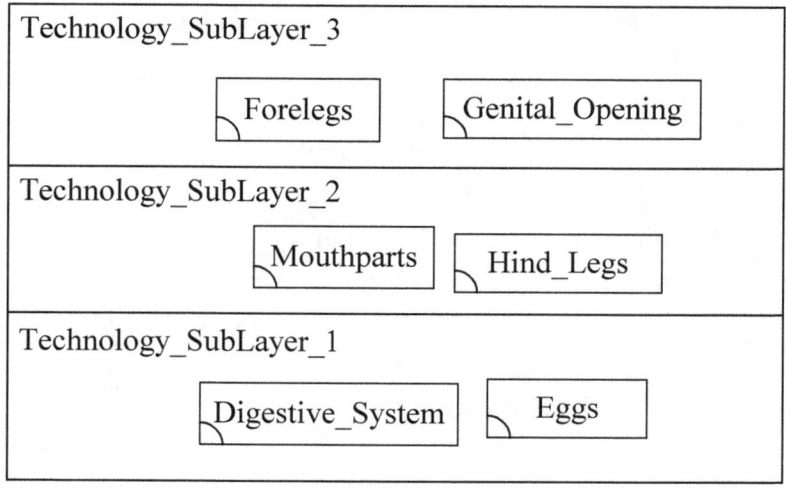

Figure 17-3   FD of the *Female Mantis*

17-3 Component Operation Diagram

We use a component operation diagram (COD) to define the operations of all components of the *female mantis* as shown in Figure 17-4. COD is the third fundamental diagram to achieve structure-behavior coalescence. In the figure, component *Forelegs* has two operations: *Grip* and *Moved_forward*; component *Genital_Opening* has one operation: *Copulate*; component *Mouthparts* has one operation: *Chew*; component *Hind_Legs* has one operation: *Brought_forward*; component *Digestive_System* has one operation: *Digest*; component *Eggs* has one operation: *Inseminate*.

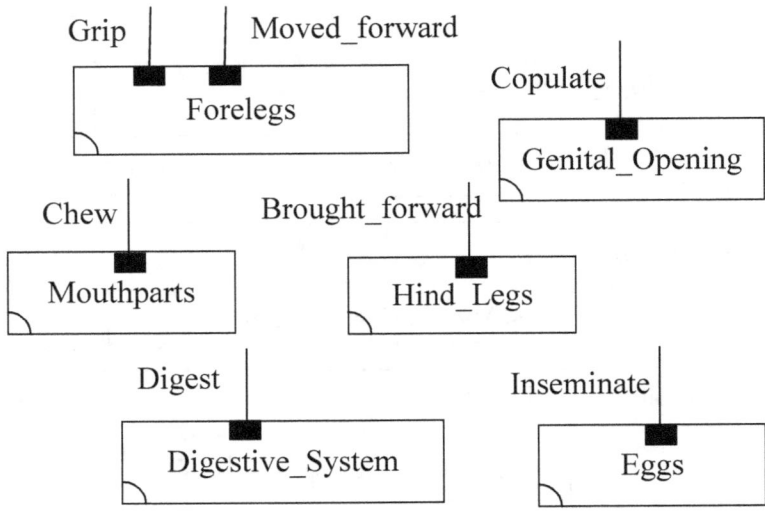

Figure 17-4   COD of the *Female Mantis*

17-4 Component Connection Diagram

We use a component connection diagram (CCD) to define the connections among components and actors of the *female mantis* as shown in Figure 17-5. CCD is the fourth fundamental diagram to achieve structure-behavior coalescence.

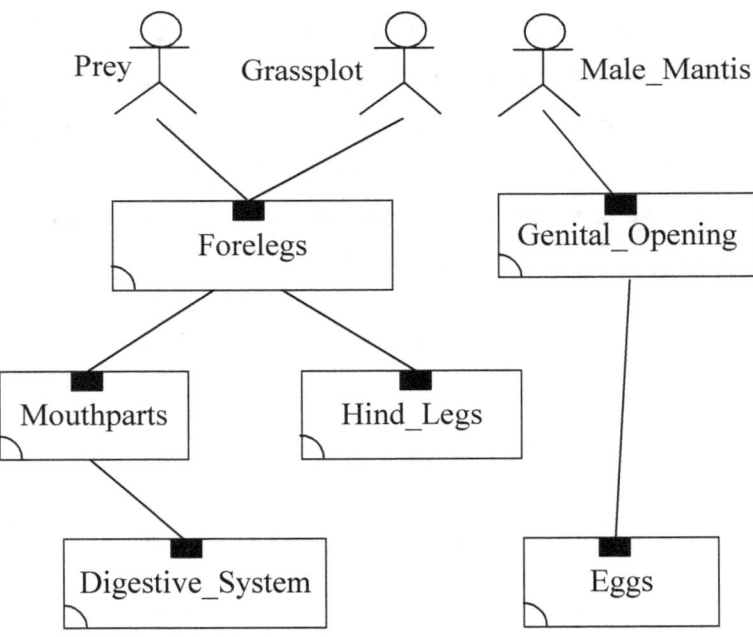

Figure 17-5   CCD of the *Female Mantis*

In Figure 17-5, actor *Prey* has a connection with the *Forelegs* component; actor *Grassplot* has a connection with the *Forelegs* component; actor *Male_Mantis* has a connection with the *Genital_Opening* component; component *Forelegs* has a connection with each one of the *Mouthparts* and *Hind_Legs* components; component *Mouthparts* has a connection with the *Digestive_System* component; component *Genital_Opening* has a connection with the *Eggs* component.

17-5 Structure-Behavior Coalescence Diagram

We use a structure-behavior coalescence diagram (SBCD) to define the systems structure and systems behavior coexisting in the *female mantis* as shown in Figure 17-6. SBCD is the fifth fundamental diagram to achieve structure-behavior coalescence. In the figure, interactions among the *Prey*, *Grassplot*, *Male_Mantis* actors and the *Forelegs*, *Genital_Opening*, *Mouthparts*, *Hind_Legs*, *Digestive_System*, *Eggs* components shall draw forth the *Preying*, *Walking* and *Reproducing* behaviors.

160

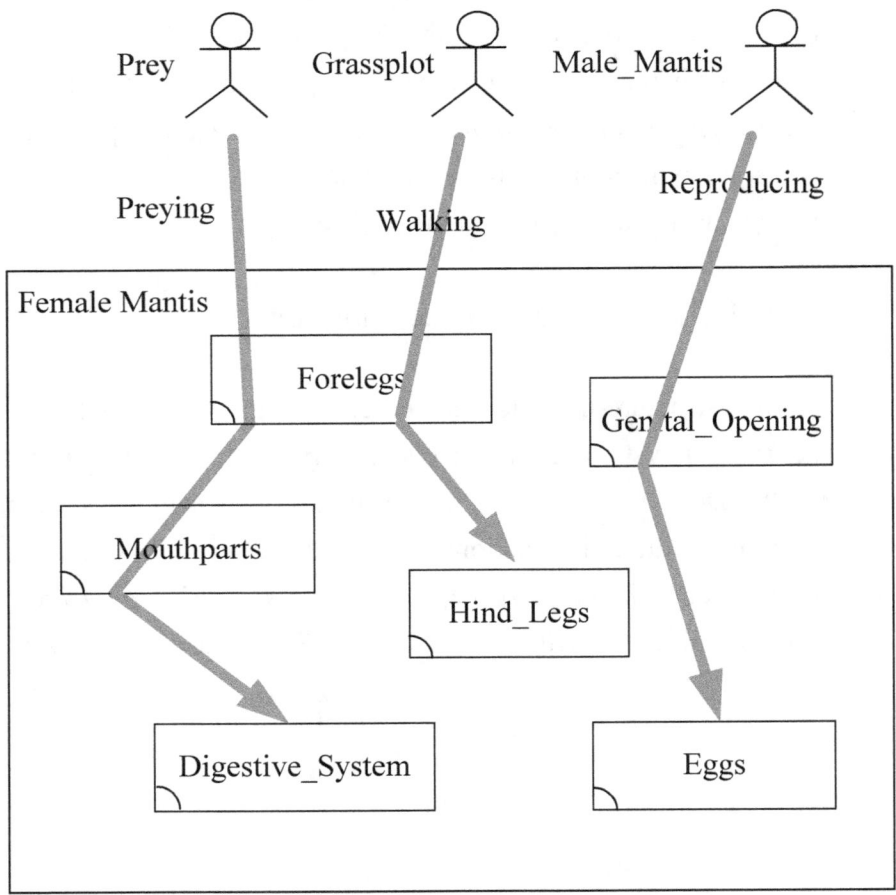

Figure 17-6   SBCD of the *Female Mantis*

The overall behavior of a system is the aggregation of all its individual behaviors. For example, the overall behavior of the *female mantis* includes the *Preying*, *Walking* and *Reproducing* behaviors.

Be noticed that the *Preying*, *Walking* and *Reproducing* behaviors are mutually independent of each other. They tend to be executed concurrently [Hoar85, Miln89, Miln99].

The major purpose of SBC architecture description language is to define an integrated whole of a system. In Figure 17-6, we not only see its structure, but also see at the same time its behavior in the *female mantis'* SBCD

17-6 Interaction Flow Diagram

The overall behavior of the *female mantis* includes three individual behaviors: *Preying*, *Walking* and *Reproducing*. Each individual behavior is represented by an execution path. We use an interaction flow diagram (IFD) to define each one of these execution paths. IFD is the sixth fundamental diagram to achieve structure-behavior coalescence.

Figure 17-7 shows an IFD of the *Preying* behavior. First, actor *Prey* interacts with the *Forelegs* component through the *Grip* operation call interaction. Second, component *Forelegs* interacts with the *Mouthparts* component through the *Chew* operation call interaction. Finally, component *Mouthparts* interacts with the *Digestive_System* component through the *Digest* operation call interaction.

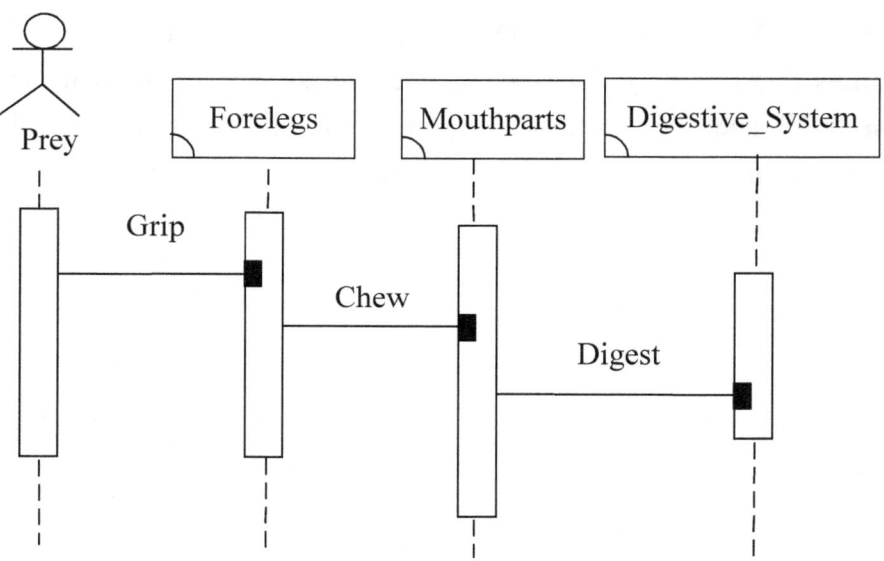

Figure 17-7   IFD of the *Preying* Behavior

Figure 17-8 shows an IFD of the *Walking* behavior. First, actor *Grassplot* interacts with the *Forelegs* component through the *Moved_forward* operation call interaction. Finally, component *Forelegs* interacts with the *Hind_Legs* component through the *Brought_forward* operation call interaction.

162

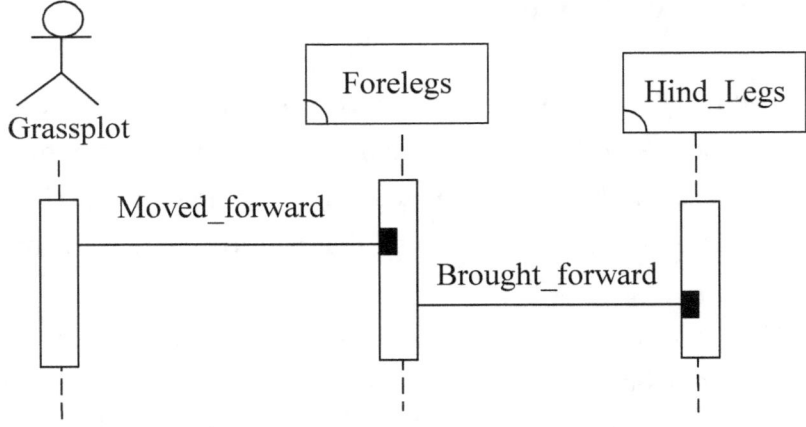

Figure 17-8   IFD of the *Walking* Behavior

Figure 17-9 shows an IFD of the *Reproducing* behavior. First, actor *Male_Mantis* interacts with the *Genital_Opening* component through the *Copulate* operation call interaction. Finally, component *Genital_Opening* interacts with the *Eggs* component through the *Inseminate* operation call interaction.

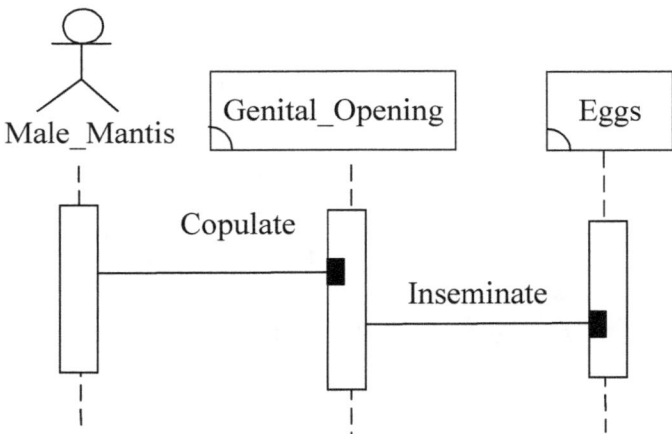

Figure 17-9   IFD of the *Reproducing* Behavior

Chapter 18: Systems Thinking 2.0 Defining the Human Body

This chapter demonstrates how to achieve systems thinking 2.0 (architectural thinking) defining the *human body*, through the application of SBC architecture description language (SBC-ADL).

Generally, the overall behavior of the *human body* is prominently represented by three individual behaviors: *Eating*, *Respiring* and *Combating* as shown in Figure 18-1.

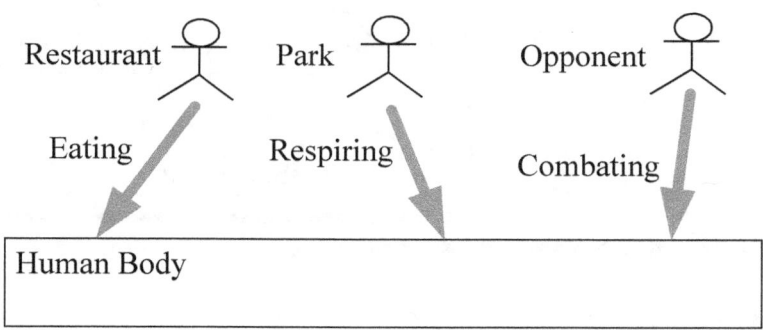

Figure 18-1   Three Behaviors of the *Human Body*

Using the SBC architecture description language, we shall go through: a) architecture hierarchy diagram, b) framework diagram, c) component operation diagram, d) component connection diagram, e) structure-behavior coalescence diagram and f) interaction flow diagram, to accomplish systems thinking 2.0 (architectural thinking) defining the *human body*.

18-1 Architecture Hierarchy Diagram

We use an architecture hierarchy diagram (AHD) to define the multi-level composition and decomposition of the *human body*. AHD is the first fundamental diagram to achieve structure-behavior coalescence. As shown in Figure 18-2, *Human Body* is composed of *Mouth*, *Nose*, *Hand* and *Subsystem_2*; *Subsystem_2* is composed of *Stomach*, *Intestine*, *Lung* and *Subsystem_1*; *Subsystem_1* is composed of *Blood_Vessel* and *Cells*.

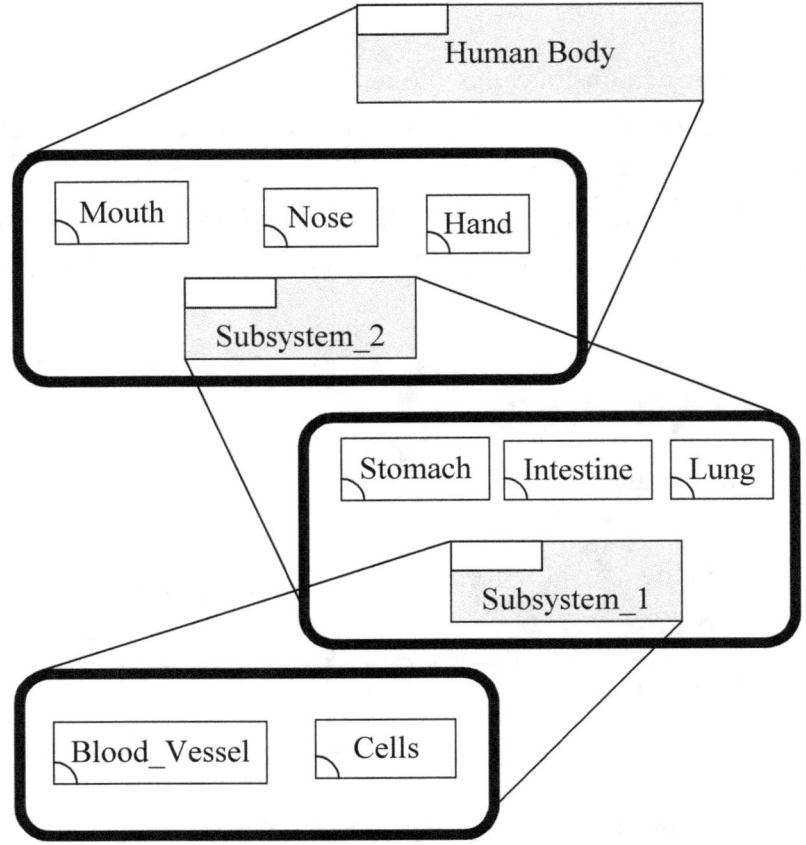

Figure 18-2   AHD of the *Human Body*

In Figure 18-2, *Human Body*, *Subsystem_2* and *Subsystem_1* are aggregated systems while *Mouth*, *Nose*, *Hand*, *Stomach*, *Intestine*, *Lung*, *Blood_Vessel* and *Cells* are non-aggregated systems.

18-2 Framework Diagram

We use a framework diagram (FD) to define the multi-layer composition and decomposition of the *human body* as shown in Figure 18-3. FD is the second fundamental diagram to achieve structure-behavior coalescence. In the figure, *Technology_SubLayer_3* contains the *Mouth*, *Nose* and *Hand* components; *Technology_SubLayer_2* contains the *Stomach*, *Intestine* and *Lung* components; *Technology_SubLayer_1* contains the *Blood_Vessel* and *Cells* components.

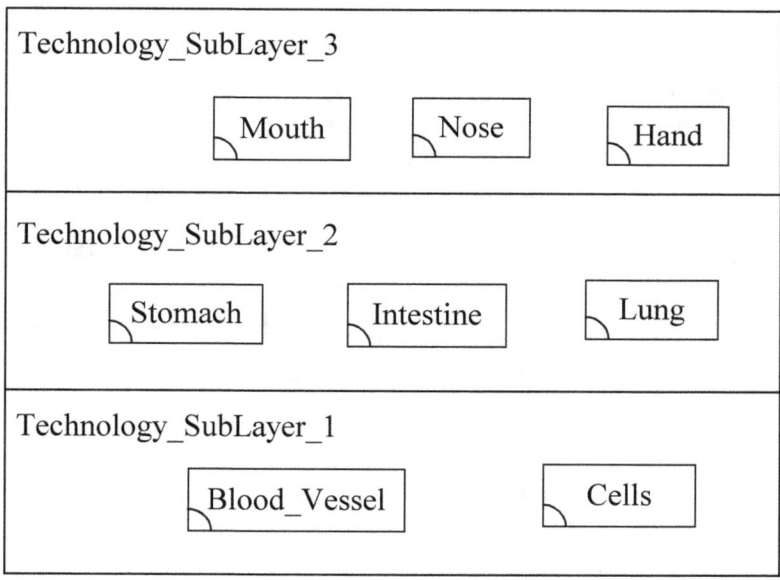

Figure 18-3   FD of the *Human Body*

18-3 Component Operation Diagram

We use a component operation diagram (COD) to define the operations of all components of the *human body* as shown in Figure 18-4. COD is the third fundamental diagram to achieve structure-behavior coalescence. In the figure, component *Mouth* has one operation: *Chew*; component *Nose* has one operation: *Breathe*; component *Stomach* has one operation: *Digest*; component *Lung* has one operation: *Exchange_gas*; component *Hand* has one operation: *Punch*; component *Intestine* has one operation: *Absorb_nutrients*; component *Blood_Vessel* has two operations: *Transport_nutrients* and *Transport_gas*; component *Eggs* has three operations: *Store_nutrients*, *Respire* and *Consume_nutrients*.

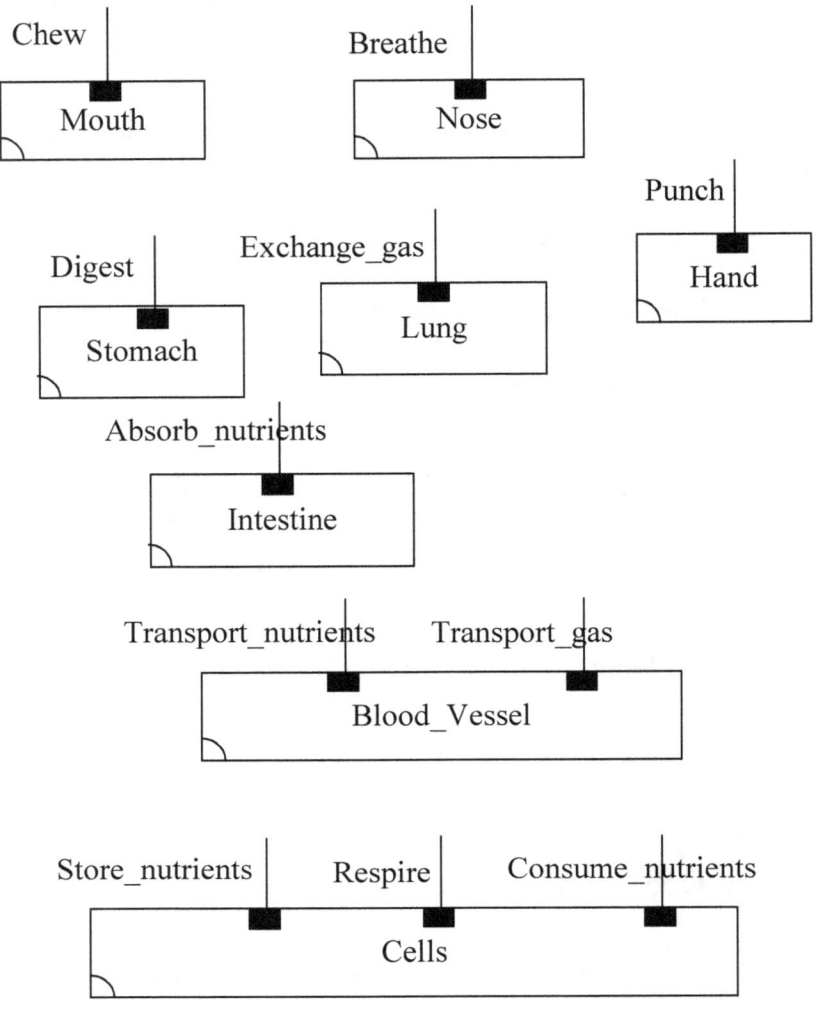

Figure 18-4   COD of the *Human Body*

18-4 Component Connection Diagram

We use a component connection diagram (CCD) to define the connections among components and actors of the *human body* as shown in Figure 18-5. CCD is the fourth fundamental diagram to achieve structure-behavior coalescence.

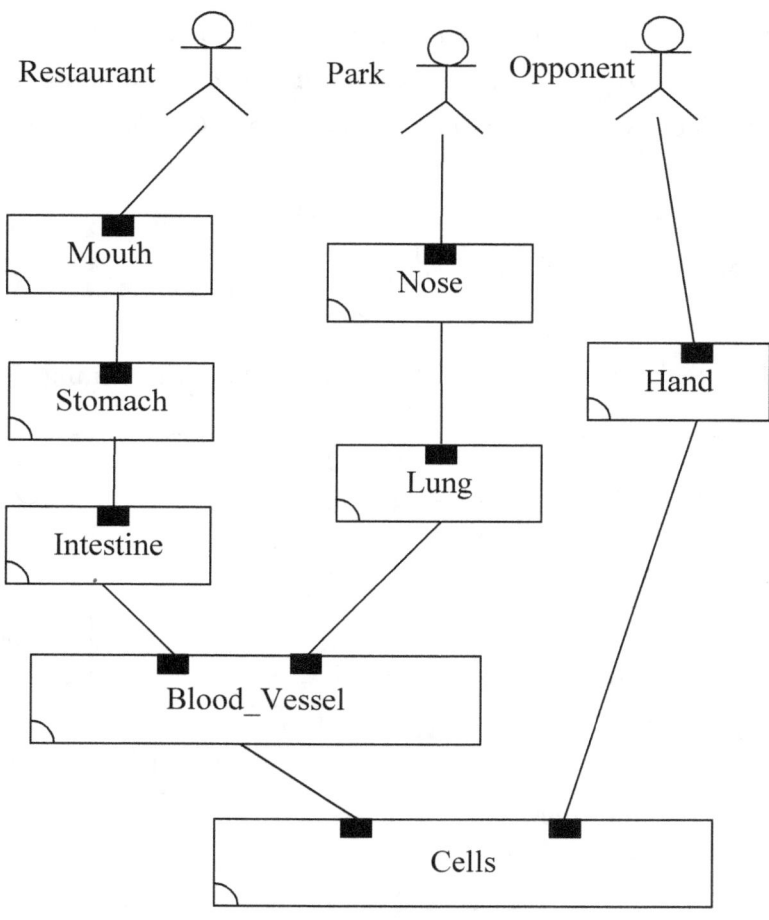

Figure 18-5   CCD of the *Human Body*

In Figure 18-5, actor *Restaurant* has a connection with the *Mouth* component; actor *Park* has a connection with the *Nose* component; actor *Opponent* has a connection with the *Hand* component; component *Mouth* has a connection with the *Stomach* component; component *Stomach* has a connection with the *Intestine* component; component *Nose* has a connection with the *Lung* component; component *Intestine* has a connection with the *Blood_Vessel* component; component *Lung* has a connection with the *Blood_Vessel* component; component *Hand* has a connection with the *Cells* component; component *Blood_Vessel* has a connection with the *Cells* components.

18-5 Structure-Behavior Coalescence Diagram

We use a structure-behavior coalescence diagram (SBCD) to define the systems structure and systems behavior coexisting in the *human body* as shown in Figure 18-6. SBCD is the fifth fundamental diagram to achieve structure-behavior coalescence. In the figure, interactions among the *Restaurant, Park, Opponent* actors and the *Mouth, Nose, Hand, Stomach, Intestine, Lung, Blood_Vessel, Cells* components shall draw forth the *Eating, Respiring* and *Combating* behaviors.

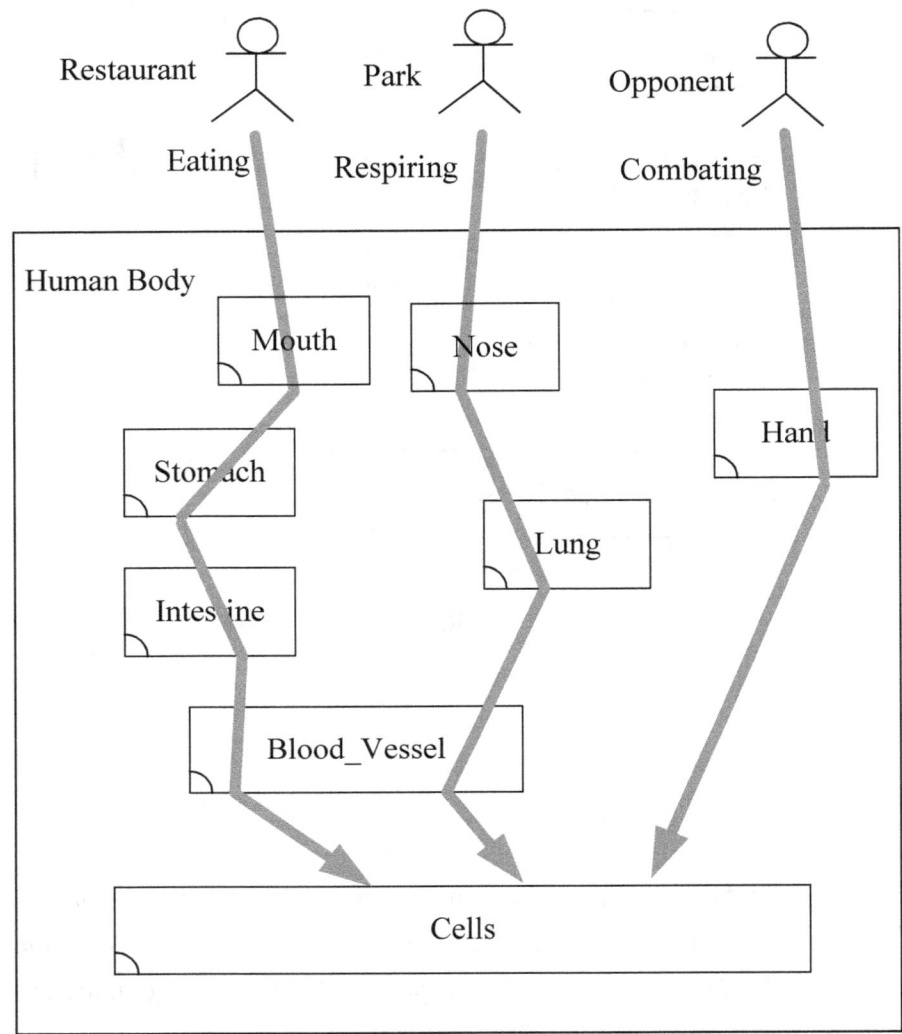

Figure 18-6   SBCD of the *Human Body*

The overall behavior of a system is the aggregation of all its individual behaviors. For example, the overall behavior of the *human body* includes the *Eating, Respiring* and *Combating* behaviors.

Be noticed that the *Eating, Respiring* and *Combating* behaviors are mutually independent of each other. They tend to be executed concurrently [Hoar85, Miln89, Miln99].

The major purpose of SBC architecture description language is to define an integrated whole of a system. In Figure 18-6, we not only see its structure, but also see at the same time its behavior in the *human body*'s SBCD.

18-6 Interaction Flow Diagram

The overall behavior of the *human body* includes three individual behaviors: *Eating*, *Respiring* and *Combating*. Each individual behavior is represented by an execution path. We use an interaction flow diagram (IFD) to define each one of these execution paths. IFD is the sixth fundamental diagram to achieve structure-behavior coalescence.

Figure 18-7 shows an IFD of the *Eating* behavior. First, actor *Restaurant* interacts with the *Mouth* component through the *Chew* operation call interaction. Second, component *Mouth* interacts with the *Stomach* component through the *Digest* operation call interaction. Third, component *Stomach* interacts with the *Intestine* component through the *Absorb_nutrients* operation call interaction. Fourth, component *Intestine* interacts with the *Blood_Vessel* component through the *Transport_nutrients* operation call interaction. Finally, component *Blood_Vessel* interacts with the *Cells* component through the *Store_nutrients* operation call interaction.

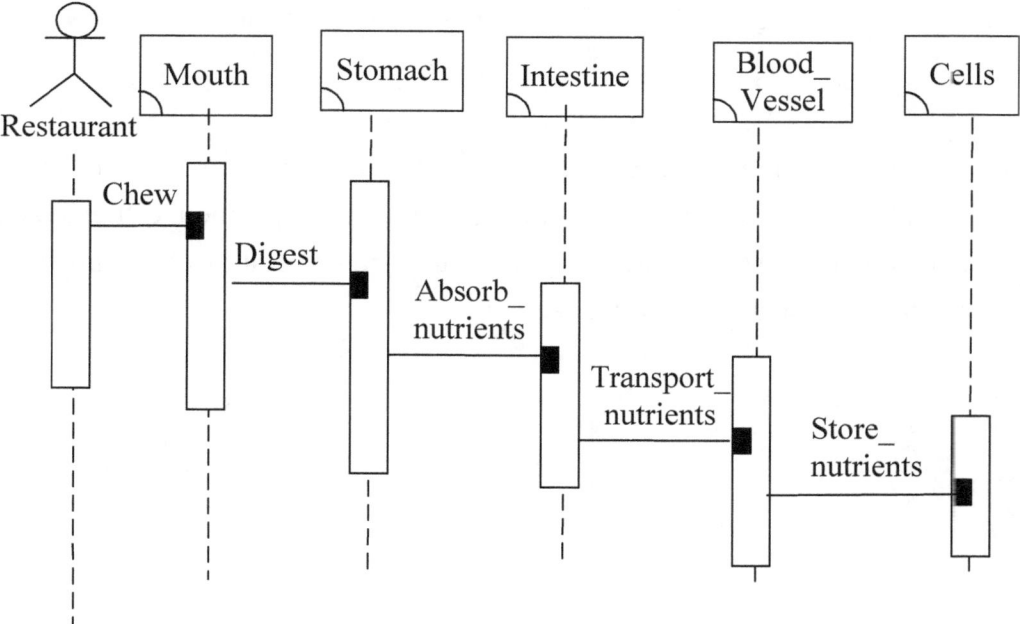

Figure 18-7   IFD of the *Eating* Behavior

Figure 18-8 shows an IFD of the *Respiring* behavior. First, actor *Park* interacts with the *Nose* component through the *Breathe* operation call interaction. Second, component *Nose* interacts with the *Lung* component through the *Exchange_gas* operation call interaction. Third, component *Lung* interacts with the *Blood_Vessel* component through the *Transport_gas* operation call interaction. Finally, component *Blood_Vessel* interacts with the *Cells* component through the *Respire* operation call interaction.

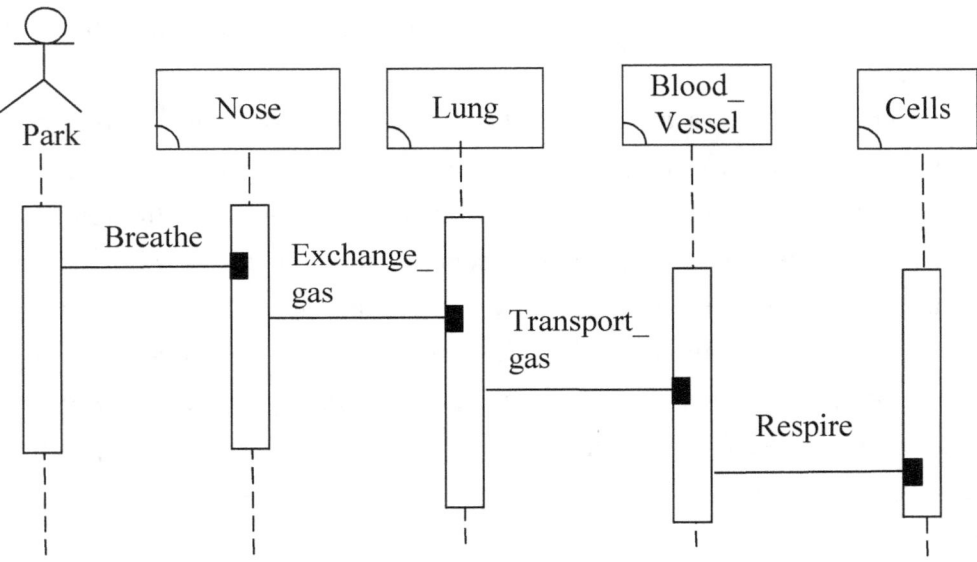

Figure 18-8   IFD of the *Respiring* Behavior

Figure 18-9 shows an IFD of the *Combating* behavior. First, actor *Opponent* interacts with the *Hand* component through the *Punch* operation call interaction. Finally, component *Hand* interacts with the component *Cells* through the *Consume_nutrients* operation call interaction.

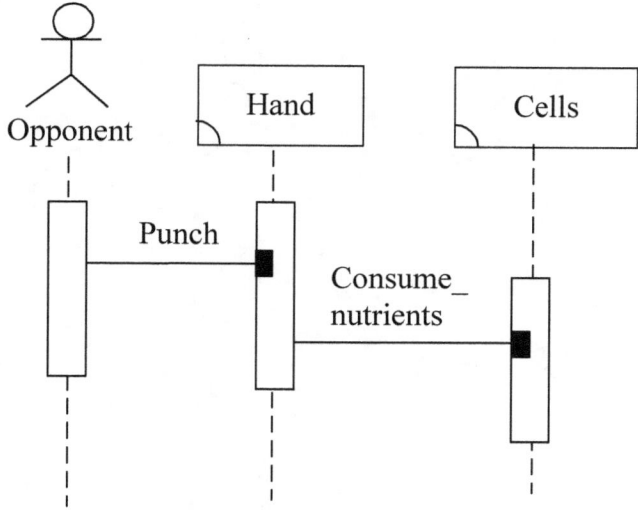

Figure 18-9   IFD of the *Combating* Behavior

Chapter 19: Systems Thinking 2.0 Defining the Disaster

This chapter demonstrates how to achieve systems thinking 2.0 (architectural thinking) defining the *disaster*, through the application of SBC architecture description language (SBC-ADL).

In general, the overall behavior of the *disaster* is prominently represented by three individual behaviors: *Flooding*, *Catching_Fire* and *Quaking* as shown in Figure 19-1.

Figure 19-1   Three Behaviors of the *Disaster*

Using the SBC architecture description language, we shall go through: a) architecture hierarchy diagram, b) framework diagram, c) component operation diagram, d) component connection diagram, e) structure-behavior coalescence diagram and f) interaction flow diagram, to accomplish systems thinking 2.0 (architectural thinking) defining the *disaster*.

19-1 Architecture Hierarchy Diagram

We use an architecture hierarchy diagram (AHD) to define the multi-level composition and decomposition of the *disaster*. AHD is the first fundamental diagram to achieve structure-behavior coalescence. As shown in Figure 19-2, *Disaster* is composed of *Dark_Cloud*, *Wire*, *Plate* and *Subsystem_2*; *Subsystem_2* is composed of *River*, *Earth_Surface* and *Subsystem_1*; *Subsystem_1* is composed of *Car*, *House* and *Mountain*.

174

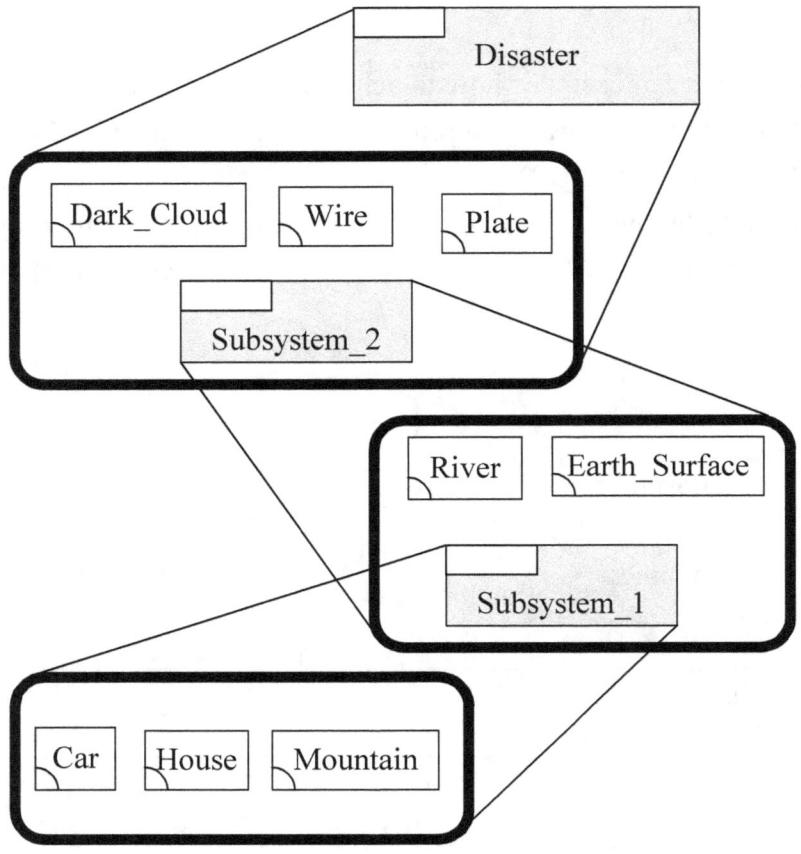

Figure 19-2   AHD of the *Disaster*

In Figure 19-2, *Disaster*, *Subsystem_2* and *Subsystem_1* are aggregated systems while *Dark_Cloud*, *Wire*, *Plate*, *River*, *Earth_Surface*, *Car*, *House* and *Mountain* are non-aggregated systems.

19-2 Framework Diagram

We use a framework diagram (FD) to define the multi-layer composition and decomposition of the *disaster* as shown in Figure 19-3. FD is the second fundamental diagram to achieve structure-behavior coalescence. In the figure, *Technology_SubLayer_3* contains the *Dark_Cloud*, *Wire* and *Plate* components; *Technology_SubLayer_2* contains the *River* and *Earth_Surface* components; *Technology_SubLayer_1* contains the *Car*, *House* and *Mountain* components.

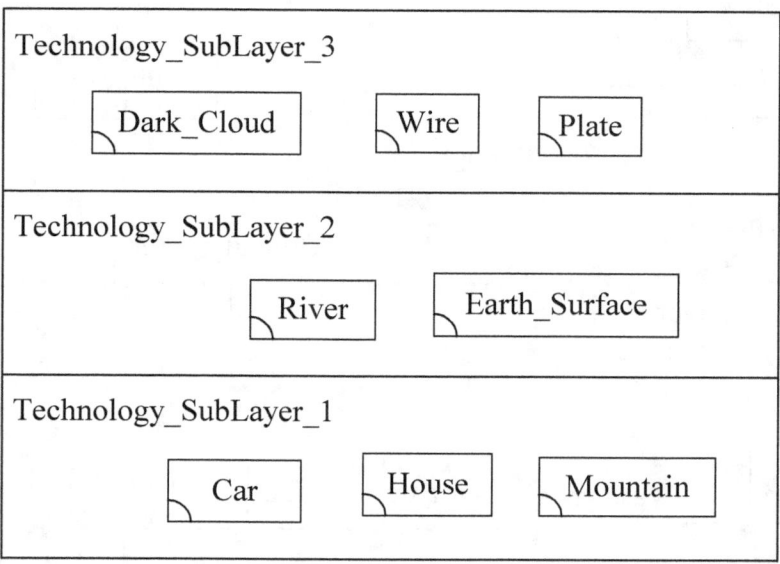

Figure 19-3   FD of the *Disaster*

19-3 Component Operation Diagram

We use the component operation diagram (COD) to define the operations of all components of the *disaster* as shown in Figure 19-4. COD is the third fundamental diagram to achieve structure-behavior coalescence. In the figure, component *Dark_Cloud* has one operation: *Heavy_rain*; component *Wire* has one operation: *Bite_through*; component *Plate* has one operation: *Move*; component *River* has one operation: *Overflow*; component *Earth_Surface* has one operation *Ground_rupture*; component *Car* has one operation: *Soak*; component *House* has three operations: *Wash_away*, *Burn_down* and *Tilt*; component *Mountain* has one operation: *Landslide*.

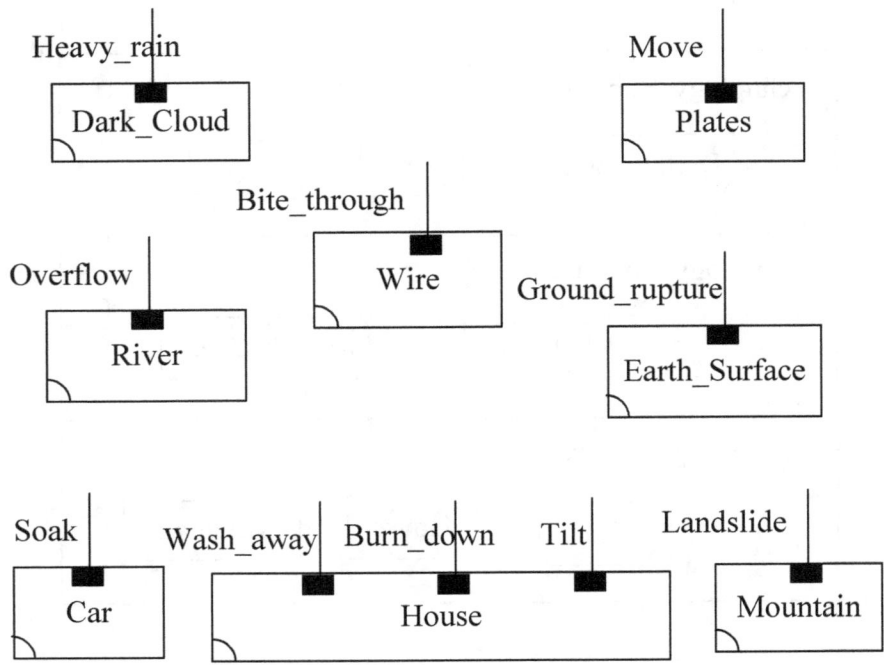

Figure 19-4   COD of the *Disaster*

19-4 Component Connection Diagram

We use a component connection diagram (CCD) to define the connections among components and actors of the *disaster* as shown in Figure 19-5. CCD is the fourth fundamental diagram to achieve structure-behavior coalescence.

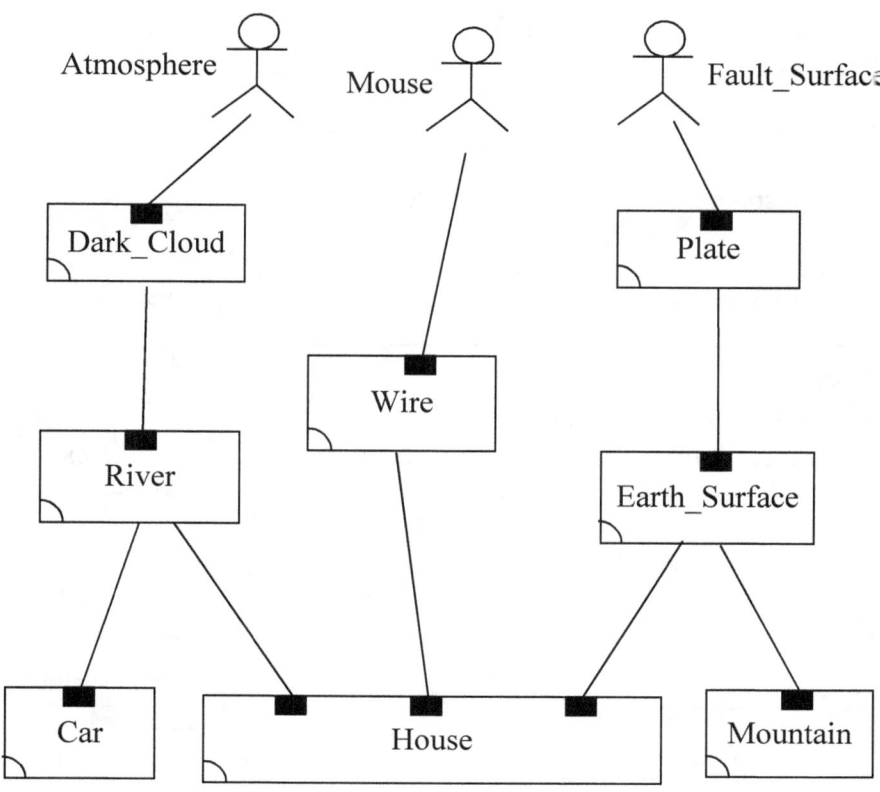

Figure 19-5   CCD of the *Disaster*

In Figure 19-5, actor *Atmosphere* have a connection with the *Dark_Cloud* component; actor *Mouse* has a connection with the *Wire* component; actor *Fault_Surface* has a connection with the *Plate* component; component *Dark_Cloud* has a connection with the *River* component; component *Wire* has a connection with the *House* component; component *Plate* has a connection with the *Earth_Surface* component; component *River* has a connection with each one of the *Car* and *House* components; component *Wire* has a connection with the *House* component; component *Earth_Surface* has a connection with each one of the *House* and *Mountain* components.

19-5 Structure-Behavior Coalescence Diagram

We use a structure-behavior coalescence diagram (SBCD) to define the systems structure and systems behavior coexisting in the *disaster* as shown in Figure 19-6. SBCD is the fifth fundamental diagram to achieve structure-behavior coalescence. In the figure, interactions among the *Atmosphere, Mouse, Fault_Surface* actors and the *Dark_Cloud*, *Wire*, *Plate*, *River*, *Earth_Surface*, *Car*, *House*, *Mountain* components shall draw forth the *Flooding*, *Catching_Fire* and *Quaking* behaviors.

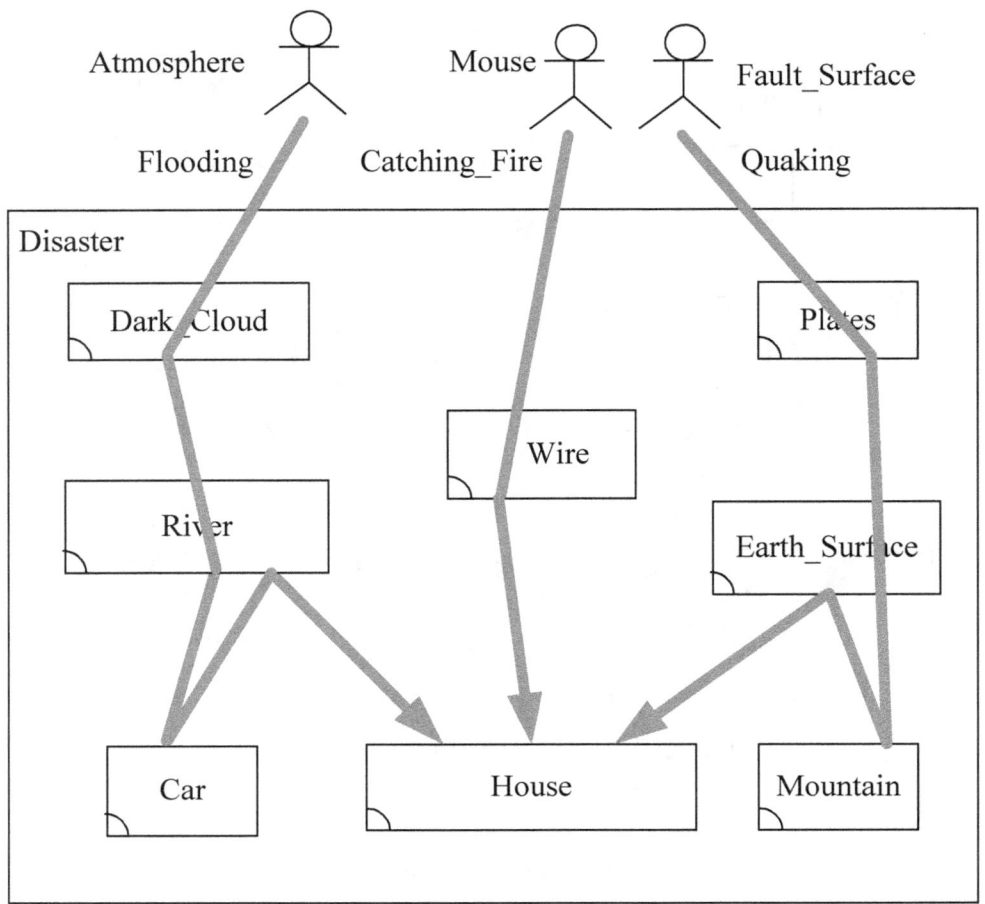

Figure 19-6   SBCD of the *Disaster*

The overall behavior of a system is the aggregation of all its individual behaviors. For example, the overall behavior of the *disaster* includes the *Flooding*, *Catching_Fire* and *Quaking* behaviors.

Be noticed that the *Flooding*, *Catching_Fire* and *Quaking* behaviors are mutually independent of each other. They tend to be executed concurrently [Hoar85, Miln89, Miln99].

The major purpose of SBC architecture description language is to define an integrated whole of a system. In Figure 19-6, we not only see its structure, but also see at the same time its behavior in the *disaster*'s SBCD.

19-6 Interaction Flow Diagram

The overall behavior of the *human body* includes three individual behaviors: *Flooding*, *Catching_Fire* and *Quaking*. Each individual behavior is represented by an execution path. We use an interaction flow diagram (IFD) to define each one of these execution paths. IFD is the sixth fundamental diagram to achieve structure-behavior coalescence.

Figure 19-7 shows an IFD of the *Flooding* behavior. First, actor *Atmosphere* interacts with the *Dark_Cloud* component through the *Heavy_rain* operation call interaction. Second, component *Dark_Cloud* interacts with the *River* component through the *Overflow* operation call interaction. Third, component *River* interacts with the *Car* component through the *Soak* operation call interaction. Finally, component *River* interacts with the *House* component through the *Wash_away* operation call interaction.

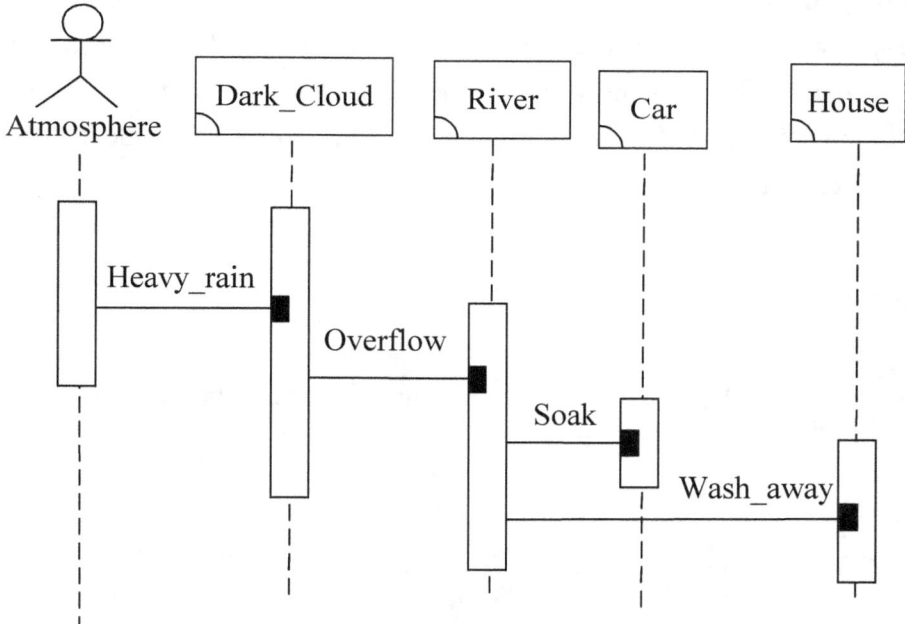

Figure 19-7   IFD of the *Flooding* Behavior

Figure 19-8 shows an IFD of the *Catching_Fire* behavior. First, actor *Mouse* interacts with the *Wire* component through the *Bite_through* operation call interaction. Finally, component *Wire* interacts with the *House* component through the *Burn_down* operation call interaction.

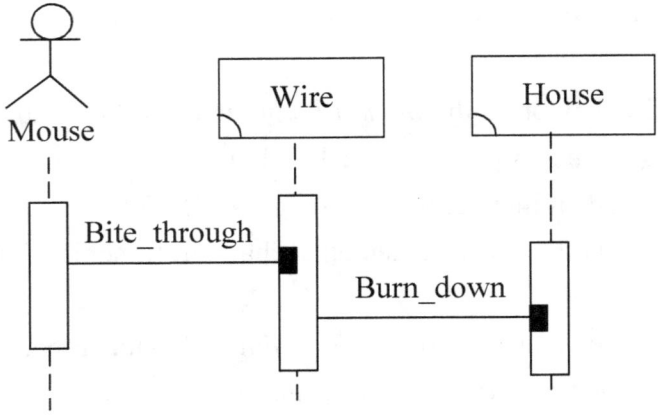

Figure 19-8   IFD of the *Catching_Fire* Behavior

Figure 19-9 shows an IFD of the *Quaking* behavior. First, actor *Fault_Surface* interacts with the *Plate* component through the *Move* operation call interaction. Second, component *Plate* interacts with the *Earth_Surface* component through the *Ground_rupture* operation call interaction. Third, component *Earth_Surface* interacts with the *Mountain* component through the *Landslide* operation call interaction. Finally, component *Earth_Surface* interacts with the *House* component through the *Tilt* operation call interaction.

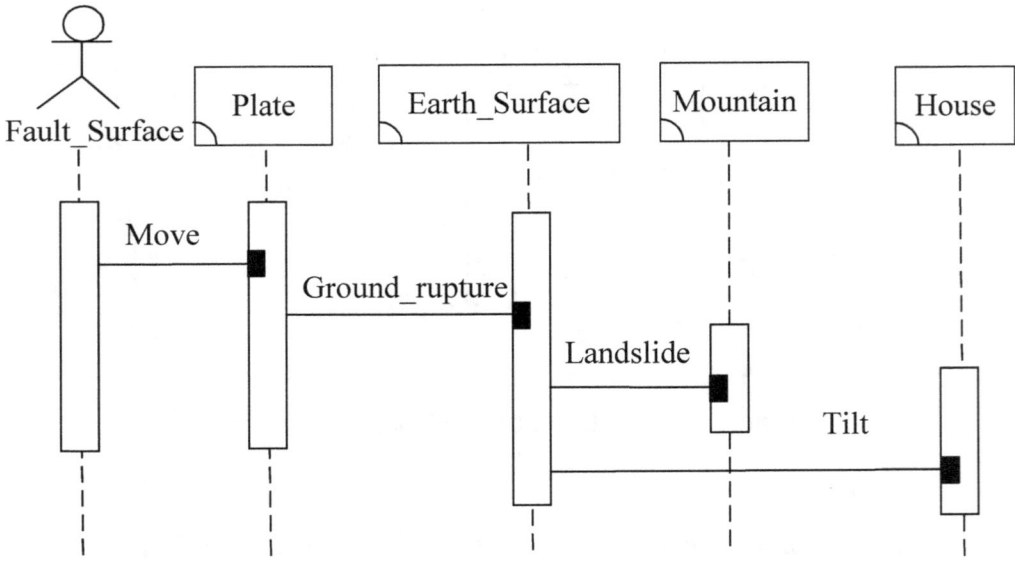

Figure 19-9   IFD of the *Quaking* Behavior

Chapter 20: Systems Thinking 2.0 Defining the Bicycle

This chapter demonstrates how to achieve systems thinking 2.0 (architectural thinking) defining the *bicycle*, through the application of SBC architecture description language (SBC-ADL).

Generally, the overall behavior of the *bicycle* is prominently represented by three individual behaviors: *Advancing*, *Turning_Left* and *Braking* as shown in Figure 20-1.

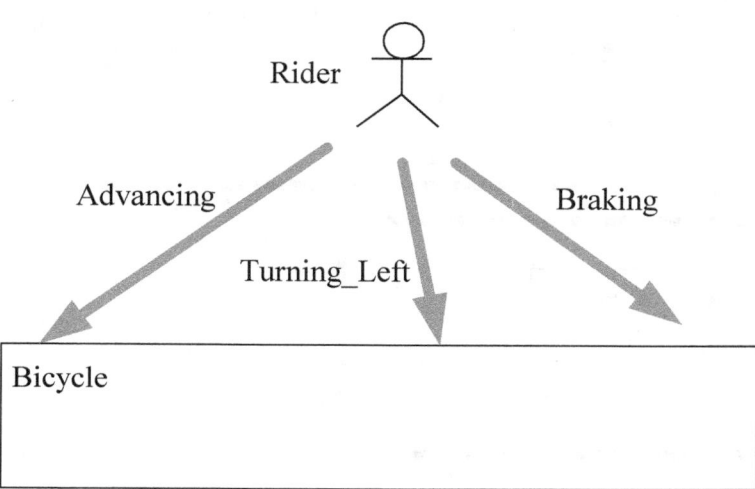

Figure 20-1   Three Behaviors of the *Bicycle*

Using the SBC architecture description language, we shall go through: a) architecture hierarchy diagram, b) framework diagram, c) component operation diagram, d) component connection diagram, e) structure-behavior coalescence diagram and f) interaction flow diagram, to accomplish systems thinking 2.0 (architectural thinking) defining the *bicycle*.

20-1 Architecture Hierarchy Diagram

We use an architecture hierarchy diagram (AHD) to define the multi-level composition and decomposition of the *bicycle*. AHD is the first fundamental diagram to achieve structure-behavior coalescence. As shown in Figure 20-2, *Bicycle* is composed of *Pedal, Handlebar, Brake_Lever* and *Subsystem_2; Subsystem_2* is composed of *Chain, Front_Wheel, Bowden_Cable* and *Subsystem_1; Subsystem_1* is composed of *Rear_Wheel* and *Friction_Pad*.

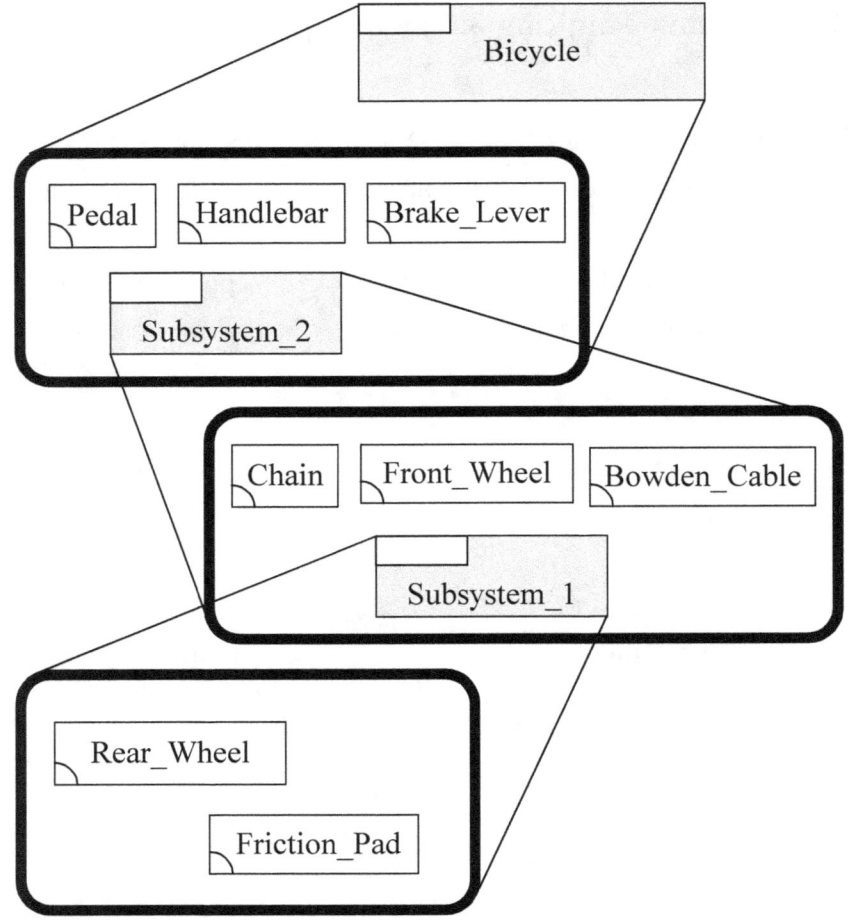

Figure 20-2   AHD of the *Bicycle*

In Figure 20-2, *Bicycle*, *Subsystem_2* and *Subsystem_1* are aggregated systems while *Pedal*, *Handlebar*, *Brake_Lever*, *Chain*, *Front_Wheel*, *Bowden_Cable*, *Rear_Wheel* and *Friction_Pad* are non-aggregated systems.

20-2 Framework Diagram

We use a framework diagram (FD) to define the multi-layer composition and decomposition of the *bicycle* as shown in Figure 20-3. FD is the second fundamental diagram to achieve structure-behavior coalescence. In the figure, *Technology_SubLayer_3* contains the *Pedal*, *Handlebar* and *Brake_Lever* components; *Technology_SubLayer_2* contains the *Chain*, *Front_Wheel* and *Bowden_Cable* components; *Technology_SubLayer_1* contains the *Rear_Wheel* and *Friction_Pad* components.

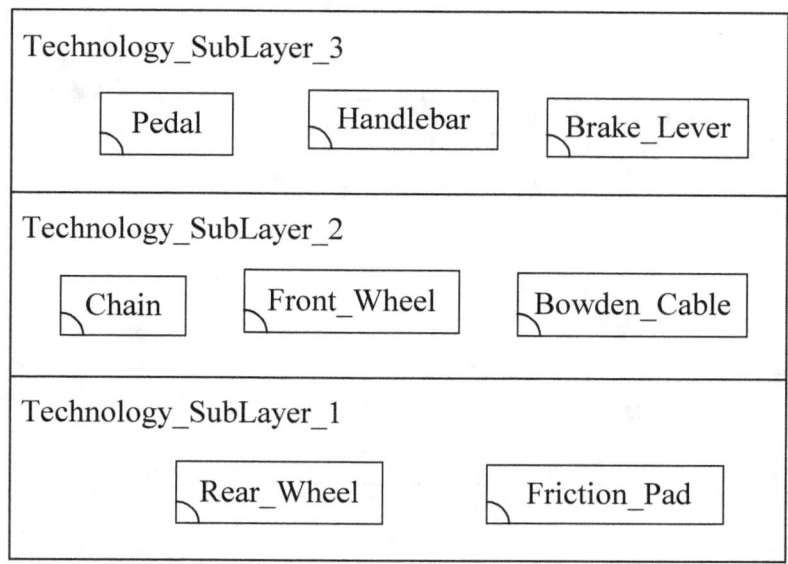

Figure 20-3   FD of the *Bicycle*

20-3 Component Operation Diagram

We use a component operation diagram (COD) to define the operations of all components of the *bicycle* as shown in Figure 20-4. COD is the third fundamental diagram to achieve structure-behavior coalescence. In the figure, component *Pedal* has one operation: *Depress*; component *Handlebar* has one operation: *Steer*; component *Brake_Lever* has one operation: *Force*; component *Chain* has one operation: *Transmit_power*; component *Front_Wheel* has one operation: *Turn_left*; component *Bowden_Cable* has one operation: *Pull*; component *Rear_Wheel* has two operations: *Roll* and *Stop_rolling*; component *Friction_Pad* has one operation: *Compress*.

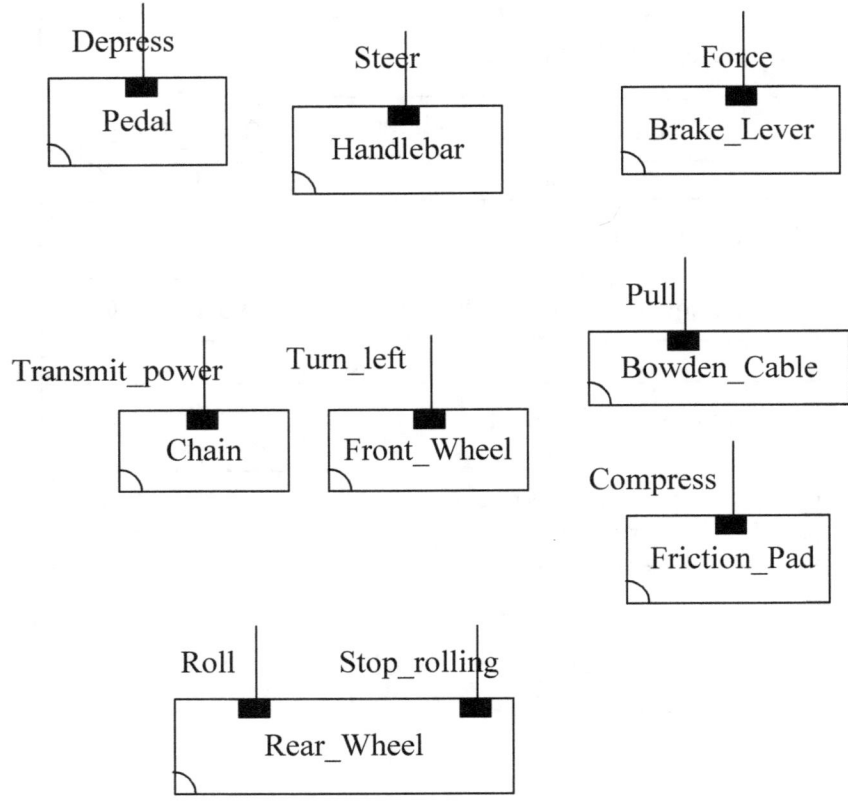

Figure 20-4   COD of the *Bicycle*

20-4 Component Connection Diagram

We use a component connection diagram (CCD) to define the connections among components and actors of the *bicycle* as shown in Figure 20-5. CCD is the fourth fundamental diagram to achieve structure-behavior coalescence.

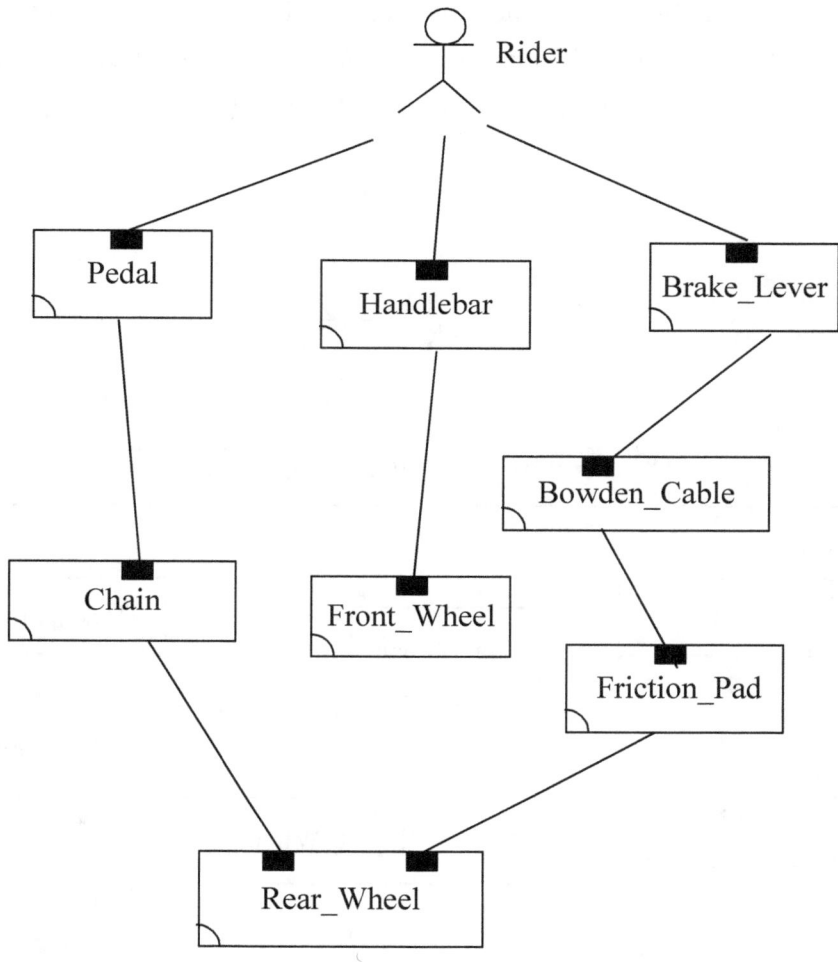

Figure 20-5   CCD of the *Bicycle*

In Figure 20-5, actor *Rider* has a connection with each one of the *Pedal*, *Handlebar* and *Brake_Lever* components; component *Pedal* has a connection with the *Chain* component; component *Handlebar* has a connection with the *Front_Wheel* component; component *Brake_Lever* has a connection with the *Bowden_Cable* component; component *Bowden_Cable* has a connection with the *Friction_Pad* component; component *Chain* has a connection with the *Rear_Wheel* component; component *Friction_Pad* has a connection with the *Rear_Wheel* component.

20-5 Structure-Behavior Coalescence Diagram

We use a structure-behavior coalescence diagram (SBCD) to define the systems structure and systems behavior coexisting in the *bicycle* as shown in Figure 20-6. SBCD is the fifth fundamental diagram to achieve structure-behavior coalescence. In the figure, interactions among the *Rider* actor and the *Pedal*, *Handlebar*, *Brake_Lever*, *Chain*, *Front_Wheel*, *Bowden_Cable*, *Rear_Wheel*, *Friction_Pad* components shall draw forth the *Advancing*, *Turning_Left* and *Braking* behaviors.

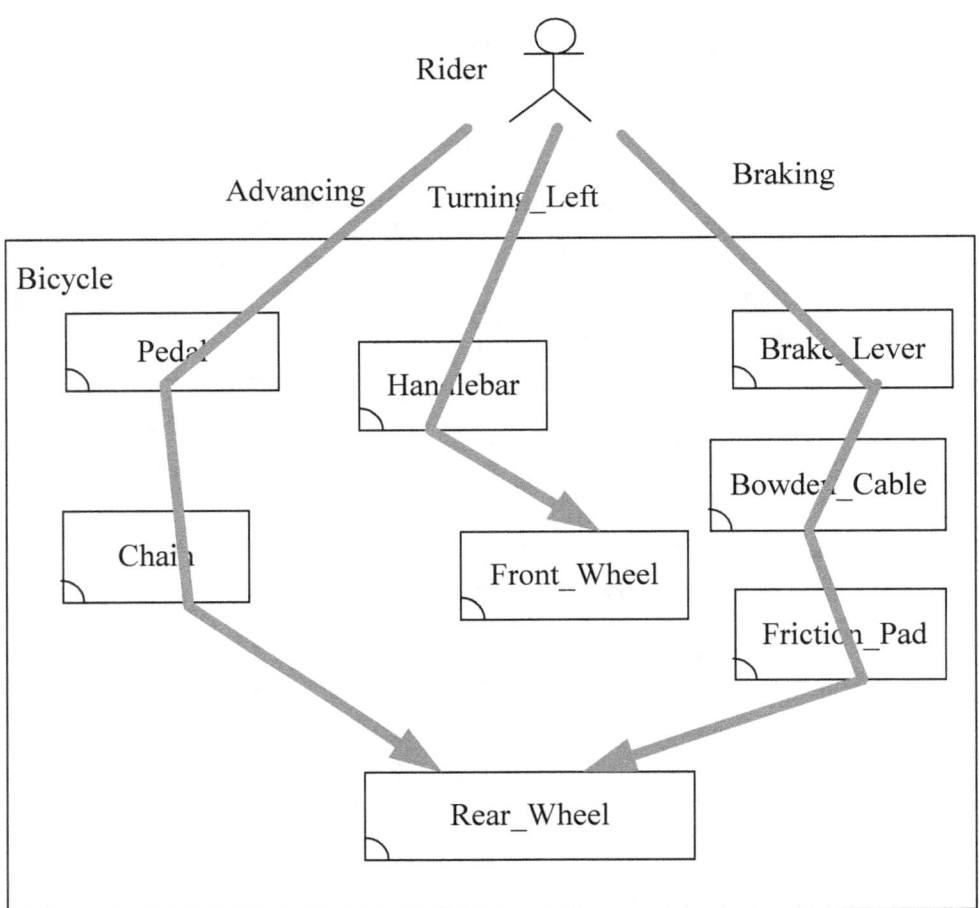

Figure 20-6   SBCD of the *Bicycle*

The overall behavior of a system is the aggregation of all its individual behaviors. For example, the overall behavior of the *bicycle* includes the *Advancing*, *Turning_Left* and *Braking* behaviors.

Be noticed that the *Advancing*, *Turning_Left* and *Braking* behaviors are mutually independent of each other. They tend to be executed concurrently [Hoar85, Miln89, Miln99].

The major purpose of SBC architecture description language is to define an integrated whole of a system. In Figure 20-6, we not only see its structure, but also see at the same time its behavior in the *bicycle*'s SBCD.

20-6 Interaction Flow Diagram

The overall behavior of the *bicycle* includes three individual behaviors: *Advancing*, *Turning_Left* and *Braking*. Each individual behavior is represented by an execution path. We use an interaction flow diagram (IFD) to define each one of these execution paths. IFD is the sixth fundamental diagram to achieve structure-behavior coalescence.

Figure 20-7 shows an IFD of the *Advancing* behavior. First, actor *Rider* interacts with the *Pedal* component through the *Depress* operation call interaction. Second, component *Pedal* interacts with the *Chain* component through the *Transmit_power* operation call interaction. Finally, component *Chain* interacts with the *Rear_Wheel* component through the *Roll* operation call interaction.

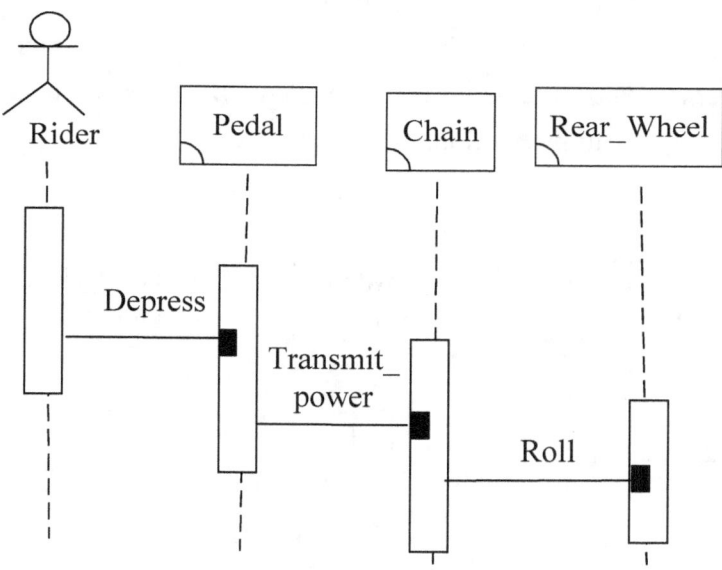

Figure 20-7   IFD of the *Advancing* Behavior

Figure 20-8 shows an IFD of the *Turning_Left* behavior. First, actor *Rider* interacts with the *Handlebar* component through the *Steer* operation call interaction. Finally, component *Handlebar* interacts with the *Front_Wheel* component through the *Turn_left* operation call interaction.

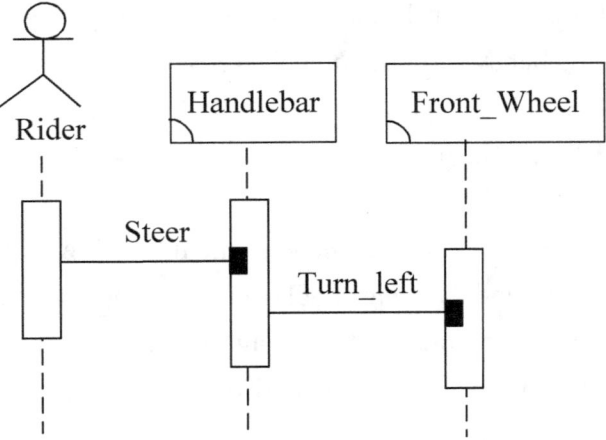

Figure 20-8   IFD of the *Turning_Left* Behavior

Figure 20-9 shows an IFD of the *Braking* behavior. First, actor *Rider* interacts with the *Brake_Lever* component through the *Force* operation call interaction. Second, component *Brake_Lever* interacts with the *Bowden_Cable* component through the *Pull* operation call interaction. Third, component *Bowden_Cable* interacts with the *Friction_Pad* component through the *Compress* operation call interaction. Finally, component *Friction_Pad* interacts with the *Rear_Wheel* component through the *Stop_rolling* operation call interaction.

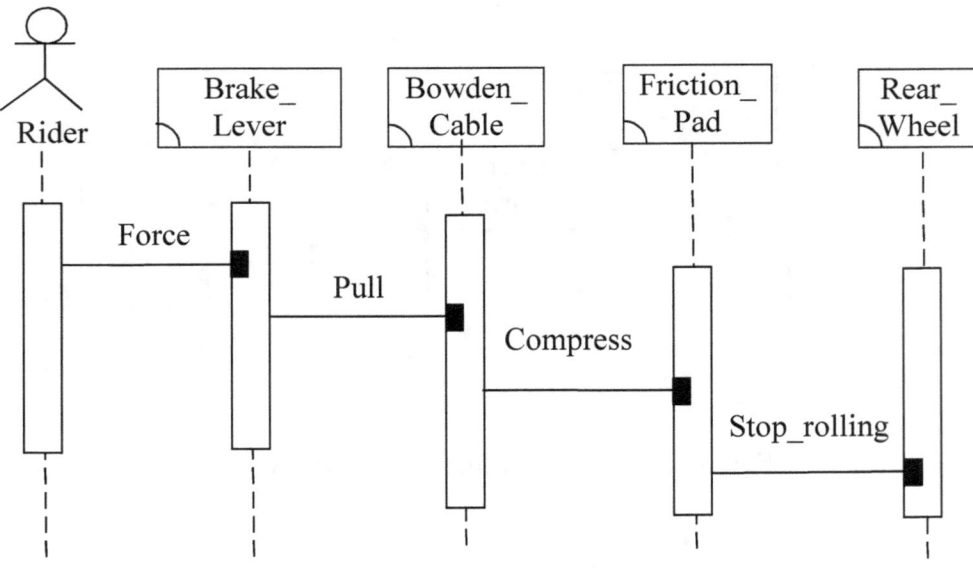

Figure 20-9   IFD of the *Braking* Behavior

Chapter 21: Systems Thinking 2.0 Defining the Purchasing Management

This chapter demonstrates how to achieve systems thinking 2.0 (architectural thinking) defining the *purchasing management*, through the application of SBC architecture description language (SBC-ADL).

Generally, the overall behavior of the *purchasing management* is prominently represented by five individual behaviors: *Purchase_Requisition*, *Quotation*, *Purchase_Order*, *Purchase* and *Collect_Supplier_Data* as shown in Figure 21-1.

Figure 21-1   Five Behaviors of the *Purchasing Management*

Using the SBC architecture description language, we shall go through: a) architecture hierarchy diagram, b) framework diagram, c) component operation diagram, d) component connection diagram, e) structure-behavior coalescence diagram and f) interaction flow diagram, to accomplish systems thinking 2.0 (architectural thinking) defining the *purchasing management*.

21-1 Architecture Hierarchy Diagram

We use an architecture hierarchy diagram (AHD) to define the multi-level composition and decomposition of the *purchasing management*. AHD is the first fundamental diagram to achieve structure-behavior coalescence. As shown in Figure 21-2, *Purchasing Management* is composed of *Purchase_Order_Coordinator*,

190

Purchasing_Department and *PM_Subsystem_2*; *PM_Subsystem_2* is composed of *Purchase_Requisition_Processing_GUI*, *Quotation_Processing_GUI*, *Purchase_Order_Processing_GUI*, *Purchase_Processing_GUI*, *Supplier_Data_Processing_GUI* and *PM_Subsystem_1*; *PM_Subsystem_1* is composed of *Purchasing_Database*.

Figure 21-2   AHD of the *Purchasing Management*

In Figure 21-2, *Purchasing Management*, *PM_Subsystem_2* and *PM_Subsystem_1* are aggregated systems while *Purchase_Order_Coordinator*, *Purchasing_Department*, *Purchase_Requisition_Processing_GUI*, *Quotation_Processing_GUI*, *Purchase_Order_Processing_GUI*, *Purchase_Processing_GUI*, *Supplier_Data_Processing_GUI* and *Purchasing_Database* are non-aggregated systems.

21-2 Framework Diagram

We use a framework diagram (FD) to define the multi-layer composition and decomposition of the *purchasing management* as shown in Figure 21-3. FD is the second fundamental diagram to achieve structure-behavior coalescence. In the figure, *Business_Layer* contains the *Purchase_Order_Coordinator* and *Purchasing_Department* components; *Application_Layer* contains the *Purchase_Requisition_Processing_GUI*, *Quotation_Processing_GUI*, *Purchase_Order_Processing_GUI*, *Purchase_Processing_GUI* and *Supplier_Data_Processing_GUI* components; *Data_Layer* contains the *Purchasing_Database* component.

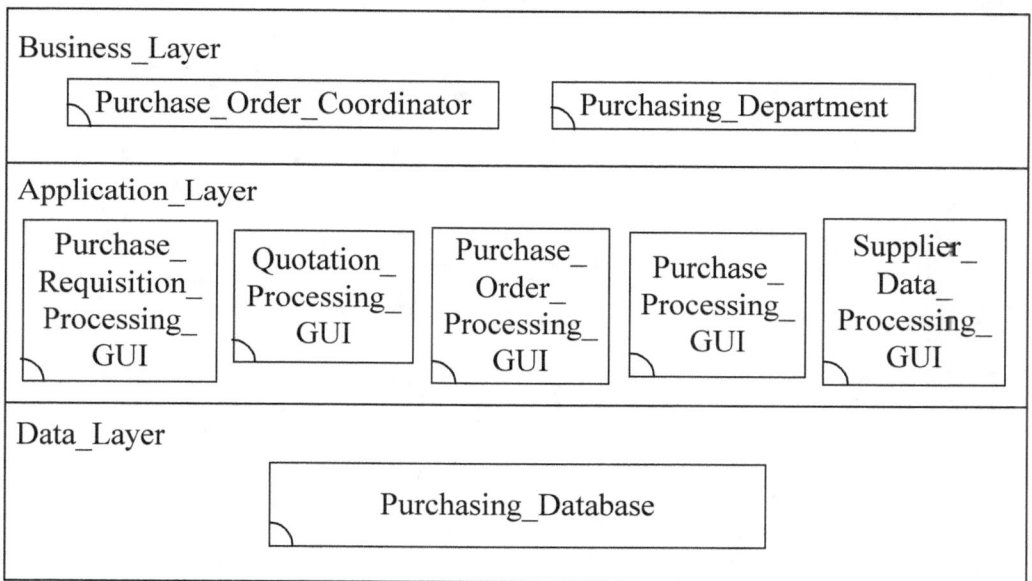

Figure 21-3   FD of the *Purchasing Management*

21-3 Component Operation Diagram

We use a component operation diagram (COD) to define the operations of all components of the *purchasing management* as shown in Figure 21-4. COD is the third fundamental diagram to achieve structure-behavior coalescence. In the figure, component *Purchase_Order_Coordinator* has one operation:

Purchase_Requisition_Verify; component *Purchasing_Department* has four operations: *Quotation_Verify*, *Purchase_Order_Verify*, *Purchase_Verify* and *Interview*; component *Purchase_Requisition_Processing_GUI* has one operation: *Purchase_Requisition_Processing_Button_Click*; component *Quotation_Processing_GUI* has one operation: *Quotation_Processing_Button_Click*; component *Purchase_Order_Processing_GUI* has one operation: *Purchase_Order_Processing_Button_Click*; component *Purchase_Processing_GUI* has one operation: *Purchase_Processing_Button_Click*; component *Supplier_Data_Processing_GUI* has one operation: *Supplier_Data_Processing_Button_Click*; component *Purchasing_Database* has five operations: *SQL_Purchase_Requisition_Insert*, *SQL_Quotation_Insert*, *SQL_Purchase_Order_Insert*, *SQL_Purchase_Insert* and *SQL_Supplier_Data_Insert*.

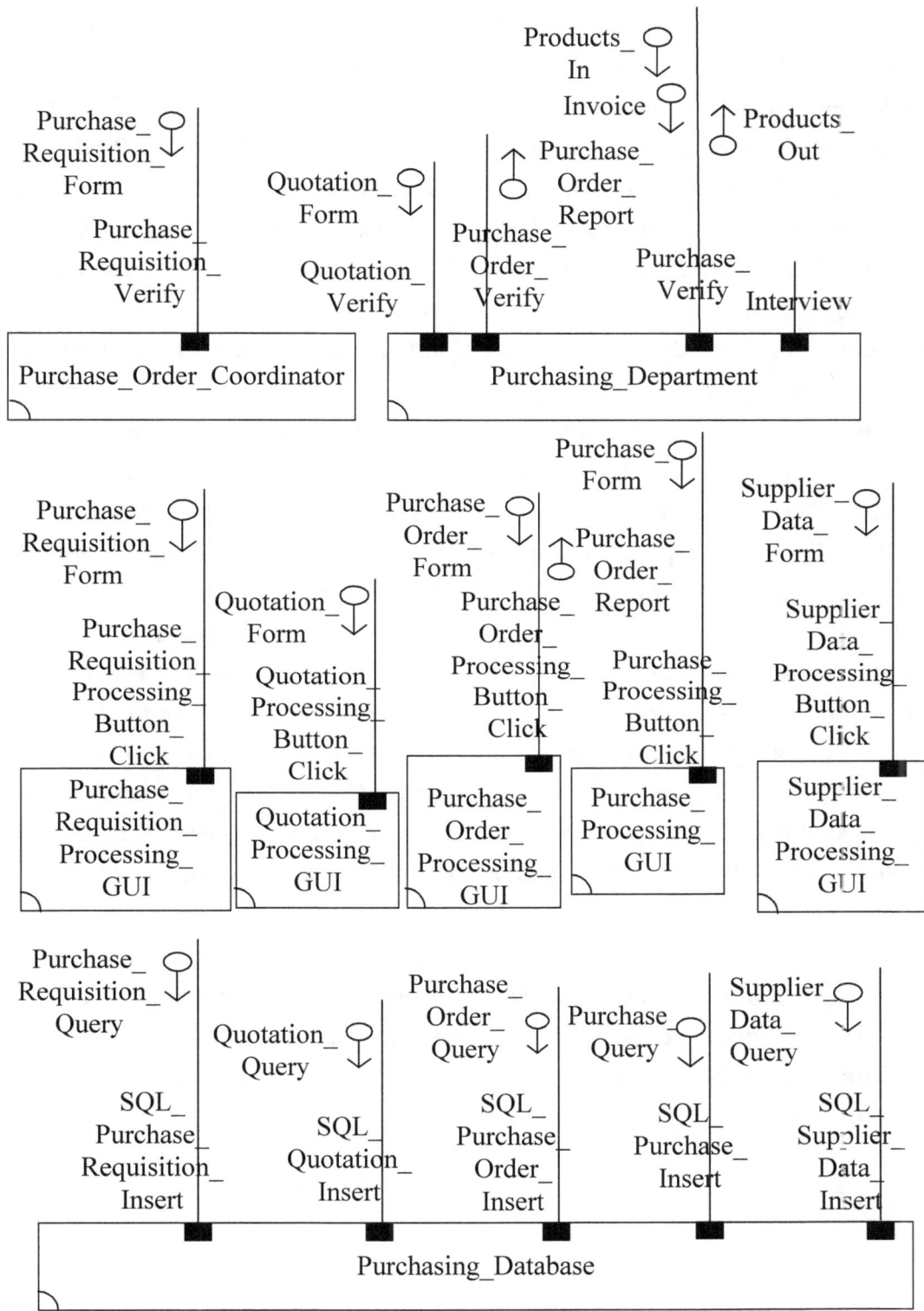

Figure 21-4   COD of the *Purchasing Management*

The operation formula of *Purchase_Requisition_Verify* is *Purchase_Requisition_Verify(In Purchase_Requisition_Form)*. The operation formula of *Quotation_Verify* is *Quotation_Verify(In Quotation_Form)*. The operation formula of *Purchase_Order_Verify* is *Purchase_Order_Verify(Out Purchase_Order_Report)*. The operation formula of *Purchase_Verify* is *Purchase_Verify(In Products_In, Invoice; Out Products_Out)*. The operation formula of *Interview* is *Interview*. The operation formula of *Purchase_Requisition_Processing_Button_Click* is *Purchase_Requisition_Processing_Button_Click(In Purchase_Requisition_Form)*. The operation formula of *Quotation_Processing_Button_Click* is *Quotation_Processing_Button_Click(In Quotation_Form)*. The operation formula of *Purchase_Order_Processing_Button_Click* is *Purchase_Order_Processing_Button_Click(In Purchase_Order_Form; Out Purchase_Order_Report)*. The operation formula of *Purchase_Processing_Button_Click* is *Purchase_Processing_Button_Click(In Purchase_Form)*. The operation formula of *Supplier_Data_Processing_Button_Click* is *Supplier_Data_Processing_Button_Click(In Supplier_Data_Form)*. The operation formula of *SQL_Purchase_Requisition_Insert* is *SQL_Purchase_Requisition_Insert(In Purchase_Requisition_Query)*. The operation formula of *SQL_Quotation_Insert* is *SQL_Quotation_Insert(In Quotation_Query)*. The operation formula of *SQL_Purchase_Order_Insert* is *SQL_Purchase_Order_Insert (In Purchase_Order_Query)*. The operation formula of *SQL_Purchase_Insert* is *SQL_Purchase_Insert (In Purchase_Query)*. The operation formula of *SQL_Supplier_Data_Insert* is *SQL_Supplier_Data_Insert(In Supplier_Data_Query)*.

Figure 21-5 shows the composite data type specification of the *Purchase_Requisition_Form* input parameter occurring in the *Purchase_Requisition_Verify(In Purchase_Requisition_Form)* and *Purchase_Requisition_Processing_Button_Click(In Purchase_Requisition_Form)* operation formulas.

Parameter	*Purchase_Requisition_Form*

Data Type	TABLE of
	  Date : Text
	  OD : Text
	  ProductNo : Text
	  Quantity : Integer
	End TABLE ;

Instances	**Purchase Requisition Form** Date: 2011/10/17 Originating_Department : Sales Dept. ProductNo Quantity __A00001(Pen)_____300_____ __A00002(Mouse)_____400_____ __A00003(Camera)_____500_____

Figure 21-5   Composite Data Type Specification

Figure 21-6 shows the composite data type specification of the *Quotation_Form* input parameter occurring in the *Quotation_Verify(In Quotation_Form)* and *Quotation_Processing_Button_Click(In Quotation_Form)* operation formulas.

Parameter	*Quotation_Form*

Data Type	TABLE of
	    Date : Text
	    SupplierName: Text
	    ProductNo : Text
	    Quantity : Integer
	    UnitPrice : Real
	    Total : Real
	End TABLE ;

Instances	**Quotation Form** Date: 2011/10/25 SupplierName : Johnson Corp. ProductNo Quantity UnitPrice A00001(Pen) 300 100.00 _A00002(Mouse)___400____200.00__ _A00003(Camera)___500____300.00__ Total : 260,000.00

Figure 21-6   Composite Data Type Specification

Figure 21-7 shows the composite data type specification of the *Purchase_Order_Report* output parameter occurring in the *Purchase_Order_Verify(Out Purchase_Order_Report)* and *Purchase_Order_Processing_Button_Click(In Purchase_Order_Form; Out Purchase_Order_Report)* operation formulas.

Parameter	*Purchase_Order_Report*

Data Type	TABLE of
	  Date : Text
	  SupplierName: Text
	  ProductNo : Text
	  Quantity : Integer
	End TABLE ;

Instances	Date : 20111118 SupplierName : Johnson Corp. <table><tr><th>ProductNo</th><th>Quantity</th></tr><tr><td>A00001(Pen)</td><td>300</td></tr><tr><td>A00002(Mouse)</td><td>400</td></tr><tr><td>A00003(Camera)</td><td>500</td></tr></table>

Figure 21-7   Composite Data Type Specification

Figure 21-8 shows the primitive data type specification of the *Products_In* input parameter occurring in the *Purchase_Verify(In Products_In, Invoice; Out Products_Out)* operation formula.

Parameter	Data Type	Instances
Products_In	Physical Object	Pen, Mouse, Camera

Figure 21-8   Primitive Data Type Specification

Figure 21-9 shows the composite data type specification of the *Invoice* input parameter occurring in the *Purchase_Verify(In Products_In, Invoice; Out Products_Out)* operation formula.

Parameter	*Invoice*
Data Type	TABLE of Date : Text SupplierName: Text ProductNo : Text Quantity : Integer UnitPrice : Real Total : Real End TABLE ;
Instances	Date : 20111130 SupplierName : Johnson Corp. ProductNo / Quantity / UnitPrice: A00001(Pen) — 300 — 100.00 A00002(Mouse) — 400 — 200.00 A00003(Camera) — 500 — 300.00 Total : 260,000.00

Figure 21-9   Composite Data Type Specification

Figure 21-10 shows the primitive data type specification of the *Products_Out* output parameter occurring in the *Purchase_Verify(In Products_In, Invoice; Out Products_Out)* operation formula.

Parameter	Data Type	Instances
Products_Out	Physical Object	Pen, Mouse, Camera

Figure 21-10   Primitive Data Type Specification

Figure 21-11 shows the composite data type specification of the *Purchase_Order_Form* input parameter occurring in the *Purchase_Order_Processing_Button_Click(In Purchase_Order_Form; Out Purchase_Order_Report)* operation formula.

Parameter	*Purchase_Order_Form*

Data Type	TABLE of
	  Date : Text
	  SupplierName: Text
	  ProductNo : Text
	  Quantity : Integer
	End TABLE ;

Instances	**Purchase Order Form** Date: 2011/11/18 SupplierName : Johnson Corp. ProductNo Quantity __A00001(Pen) ____300 ____ __A00002(Mouse)____400___ __A00003(Camera)___500____

Figure 21-11   Composite Data Type Specification

Figure 21-12 shows the composite data type specification of the *Purchase_Form* input parameter occurring in the *Purchase_Processing_Button_Click(In Purchase_Form)* operation formula.

Parameter	*Purchase_Form*
Data Type	TABLE of 　Date : Text 　SupplierName: Text 　ProductNo : Text 　Quantity : Integer 　UnitPrice : Real 　ReturnQuantity : Integer 　Total : Real End TABLE ;
Instances	**Purchase Form** 　　　　　Date: 2011/12/12 SupplierName : Johnson Corp. ProductNo　Quantity　UnitPrice　ReturnQuantity A00001(Pen)　　300　　100.00　　　　0 A00002(Mouse) 390　　200.00　　　 10 A00003(Camera)500　300.00　　　　0 　　　　　　　　　　Total : 258,000.00

Figure 21-12　　Composite Data Type Specification

Figure 21-13 shows the composite data type specification of the *Supplier_Data_Form* input parameter occurring in the *Supplier_Data_Processing_Button_Click(In Supplier_Data_Form)* operation formula.

Parameter	*Supplier_Data_Form*
Data Type	TABLE of SupplierName:Text Address :Text PhoneNumber:Text FaxNumber:Text E-mail : Text Rank : Text End TABLE ;
Instances	**Supplier Data Form** SupplierName : Johnson Corp. Address : 1232 Fair Circle, Austin, TX PhoneNumber : 512-463-8472 FaxNumber : 512-463-8499 E-mail : Johnson1122@gmail.com Rank : B

Figure 21-13   Composite Data Type Specification

Figure 21-14 shows the composite data type specification of the *Purchase_Requisition_Query* input parameter occurring in the *SQL_Purchase_Requisition_Insert(In   Purchase_Requisition_Query)* operation formula.

Parameter	*Purchase_Requisition_Query*
Data Type	TABLE of Date : Text OD : Text ProductNo : Text Quantity : Integer End TABLE ;
Instances	<table><tr><td>Date</td><td colspan="2">Originating_Department :</td></tr><tr><td>20111017</td><td colspan="2">Sales Dept.</td></tr><tr><td>ProductNo</td><td>Quantity</td></tr><tr><td>A00001(Pen)</td><td>300</td></tr><tr><td>A00002(Mouse)</td><td>400</td></tr><tr><td>A00003(Camera)</td><td>500</td></tr></table>

Figure 21-14   Composite Data Type Specification

Figure 21-15 shows the composite data type specification of the *Quotation_Query* input parameter occurring in the *SQL_Quotation_Insert(In Quotation_Query)* operation formula.

Parameter	*Quotation_Query*

Data Type	TABLE of
	  Date : Text
	  SupplierName: Text
	  ProductNo : Text
	  Quantity : Integer
	  UnitPrice : Real
	  Total : Real
	End TABLE ;

Instances	

Date	SupplierName	Total
20111025	Johnson Corp.	260,000.00

ProductNo	Quantity	UnitPrice
A00001(Pen)	300	100.00
A00002(Mouse)	400	200.00
A00003(Camera)	500	300.00

Figure 21-15   Composite Data Type Specification

Figure 21-16 shows the composite data type specification of the *Purchase_Order_Query* input parameter occurring in the *SQL_Purchase_Order_Insert(In Purchase_Order_Query)* operation formula.

Parameter	*Purchase_Order_Query*

Data Type	TABLE of
	  Date : Text
	  SupplierName: Text
	  ProductNo : Text
	  Quantity : Integer
	End TABLE ;

Instances	 | Date	SupplierName
20111118	Johnson Corp. | | ProductNo	Quantity
A00001(Pen)	300
A00002(Mouse)	400
A00003(Camera)	500	

Figure 21-16   Composite Data Type Specification

Figure 21-17 shows the composite data type specification of the *Purchase_Query* input parameter occurring in the *SQL_Purchase_Insert (In Purchase_Query)* operation formula.

Parameter	*Purchase_Query*
Data Type	TABLE of Date : Text SupplierName: Text ProductNo : Text Quantity : Integer UnitPrice : Real ReturnQuantity : Integer Total : Real End TABLE ;
Instances	(see tables below)

Date	SupplierName	Total
20111212	Johnson Corp.	258,000.00

ProductNo	Quantity	UnitPrice	ReturnQuantity
A00001(Pen)	300	100.00	0
A00002(Mouse)	390	200.00	10
A00003(Camera)	500	300.00	0

Figure 21-17   Composite Data Type Specification

Figure 21-18 shows the composite data type specification of the *Supplier_Data_Query* input parameter occurring in the *SQL_Supplier_Data_Insert(In Supplier_Data_Query)* operation formula.

Parameter	*Supplier_Data_Query*
Data Type	TABLE of SupplierName:Text Address :Text PhoneNumber:Text FaxNumber:Text E-mail : Text Rank : Text End TABLE ;
Instances	<table><tr><td>SupplierName</td><td>Johnson Corp.</td></tr><tr><td>Address</td><td>1232 Fair Circle, Austin, TX</td></tr><tr><td>PhoneNumber</td><td>512-463-8472</td></tr><tr><td>FaxNumber</td><td>512-463-8499</td></tr><tr><td>E-mail</td><td>Johnson1122@gmail.com</td></tr><tr><td>Rank</td><td>B</td></tr></table>

Figure 21-18   Composite Data Type Specification

21-4 Component Connection Diagram

We use a component connection diagram (CCD) to define the connections among components and actors of the *purchasing management* as shown in Figure 21-19. CCD is the fourth fundamental diagram to achieve structure-behavior coalescence.

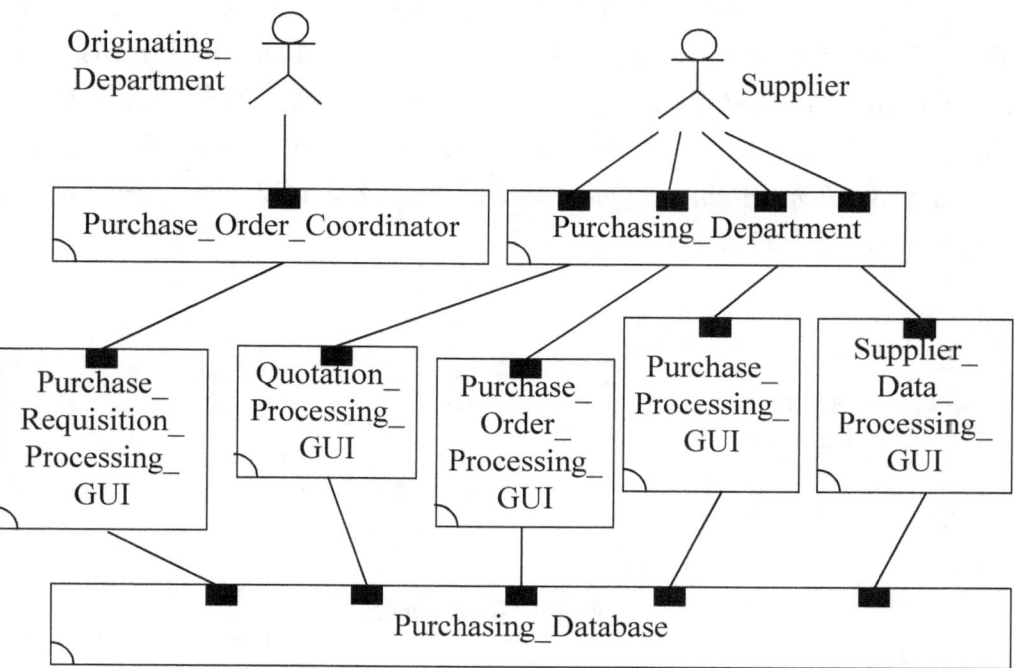

Figure 21-19   CCD of the *Purchasing Management*

In Figure 21-19, actor *Originating_Department* has a connection with the *Purchase_Order_Coordinator* component; actor *Supplier* has four connections with the *Purchasing_Department* component; component *Purchase_Order_Coordinator* has a connection with the *Purchase_Requisition_Processing_GUI* component; component *Purchasing_Department* has a connection with each one of the *Quotation_Processing_GUI*, *Purchase_Order_Processing_GUI*, *Purchase_Processing_GUI* and *Supplier_Data_Processing_GUI* components; each one of the *Purchase_Requisition_Processing_GUI*, *Quotation_Processing_GUI*, *Purchase_Order_Processing_GUI*, *Purchase_Processing_GUI* and *Supplier_Data_Processing_GUI* components has a connection with the *Purchasing_Database* component.

21-5 Structure-Behavior Coalescence Diagram

We use a structure-behavior coalescence diagram (SBCD) to define the systems structure and systems behavior coexisting in the *purchasing management* as shown in Figure 21-20. SBCD is the fifth fundamental diagram to achieve structure-behavior coalescence.

In the figure, interactions among the *Originating_Department*, *Supplier* actors and the *Purchase_Order_Coordinator*, *Purchasing_Department*, *Purchase_Requisition_Processing_GUI*, *Quotation_Processing_GUI*, *Purchase_Order_Processing_GUI*, *Purchase_Processing_GUI*, *Supplier_Data_Processing_GUI*, *Purchasing_Database* components shall draw forth the *Purchase_Requisition*, *Quotation*, *Purchase_Order*, *Purchase* and *Collect_Supplier_Data* behaviors.

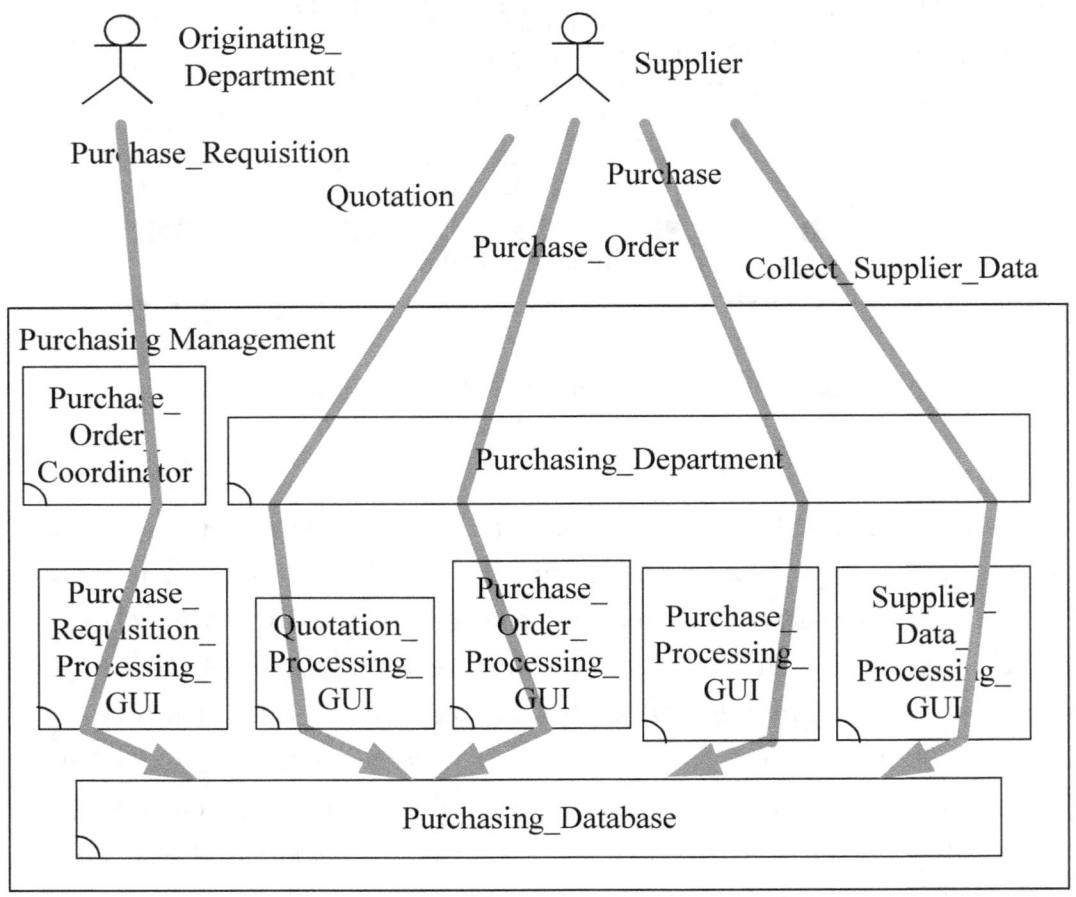

Figure 21-20   SBCD of the *Purchasing Management*

The overall behavior of a system is the aggregation of all its individual behaviors. For example, the overall behavior of the *purchasing management* includes the *Purchase_Requisition*, *Quotation*, *Purchase_Order*, *Purchase* and *Collect_Supplier_Data* behaviors.

Be noticed that the *Purchase_Requisition*, *Quotation*, *Purchase_Order*, *Purchase* and *Collect_Supplier_Data* behaviors are mutually independent of each other. They tend to be executed concurrently [Hoar85, Miln89, Miln99].

The major purpose of the architecture description language is to define an integrated whole of a system. In Figure 21-20, we not only see its structure, but also see at the same time its behavior in the *purchasing management*'s SBCD.

21-6 Interaction Flow Diagram

The overall behavior of the *purchasing management* includes five individual behaviors: *Purchase_Requisition*, *Quotation*, *Purchase_Order*, *Purchase* and *Collect_Supplier_Data*. Each individual behavior is represented by an execution path. We use an IFD to define each one of these execution paths.

Figure 21-21 shows an IFD of the *Purchase_Requisition* behavior. First, actor *Originating_Department* interacts with the *Purchase_Order_Coordinator* component through the *Purchase_Requisition_Verify* operation call interaction, carrying the *Purchase_Requisition_Form* input parameter. Next, component *Purchase_Order_Coordinator* interacts with the *Purchase_Requisition_Processing_GUI* component through the *Purchase_Requisition_Processing_Button_Click* operation call interaction, carrying the *Purchase_Requisition_Form* input parameter. Finally, component *Purchase_Requisition_Processing_GUI* interacts with the *Purchasing_Database* component through the *SQL_Purchase_Requisition_Insert* operation call interaction, carrying the *Purchase_Requisition_Query* input parameter.

Figure 21-21   IFD of the *Purchase_Requisition* Behavior

Figure 21-22 shows an IFD of the *Quotation* behavior. First, actor *Supplier* interacts with the *Purchasing_Department* component through the *Quotation_Verify* operation call interaction, carrying the *Quotation_Form* input parameter. Next, component *Purchasing_Department* interacts with the *Quotation_Processing_GUI* component through the *Quotation_Processing_Button_Click* operation call interaction, carrying the *Quotation_Form* input parameter. Finally, component *Quotation_Processing_GUI* interacts with the *Purchasing_Database* component through the *SQL_Quotation_Insert* operation call interaction, carrying the *Quotation_Query* input parameter.

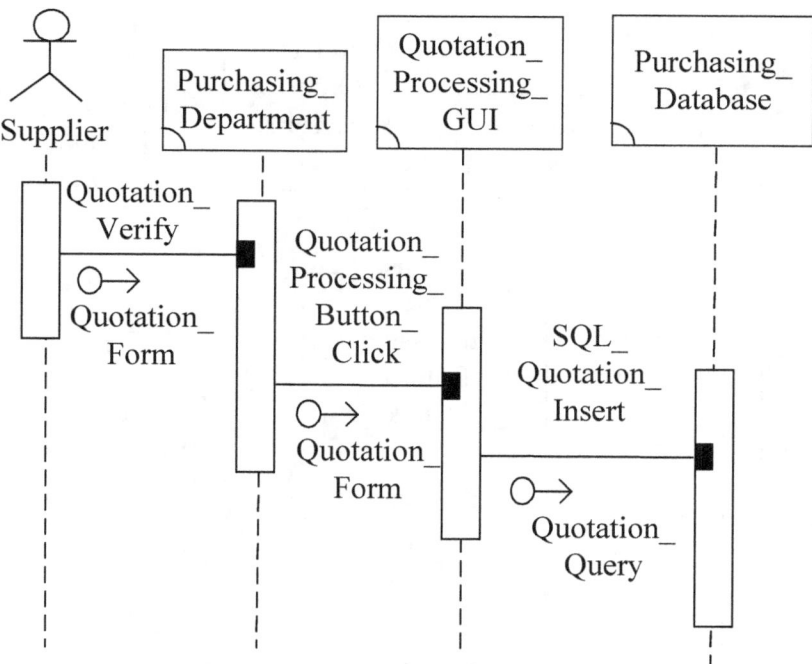

Figure 21-22   IFD of the *Quotation* Behavior

Figure 21-23 shows an IFD of the *Purchase_Order* behavior. First, actor *Supplier* interacts with the *Purchasing_Department* component through the *Purchase_Order_Verify* operation call interaction. Next, component *Purchasing_Department* interacts with the *Purchase_Order_Processing_GUI* component through the *Purchase_Order_Processing_Button_Click* operation call interaction, carrying the *Purchase_Order_Form* input parameter. Continuingly, component *Purchase_Order_Processing_GUI* interacts with the *Purchasing_Database* component through the *SQL_ Purchase_Order_Insert* operation call interaction, carrying the *Purchase_Order_Query* input parameter. Continuingly, component *Purchasing_Department* interacts with the *Purchase_Order_Processing_GUI* component through the *Purchase_Order_Processing_Button_Click* operation return interaction, carrying the *Purchase_Order_Report* output parameter. Finally, actor *Supplier* interacts with the *Purchasing_Department* component through the *Purchase_Order_Verify* operation return interaction, carrying the *Purchase_Order_Report* output parameter.

212

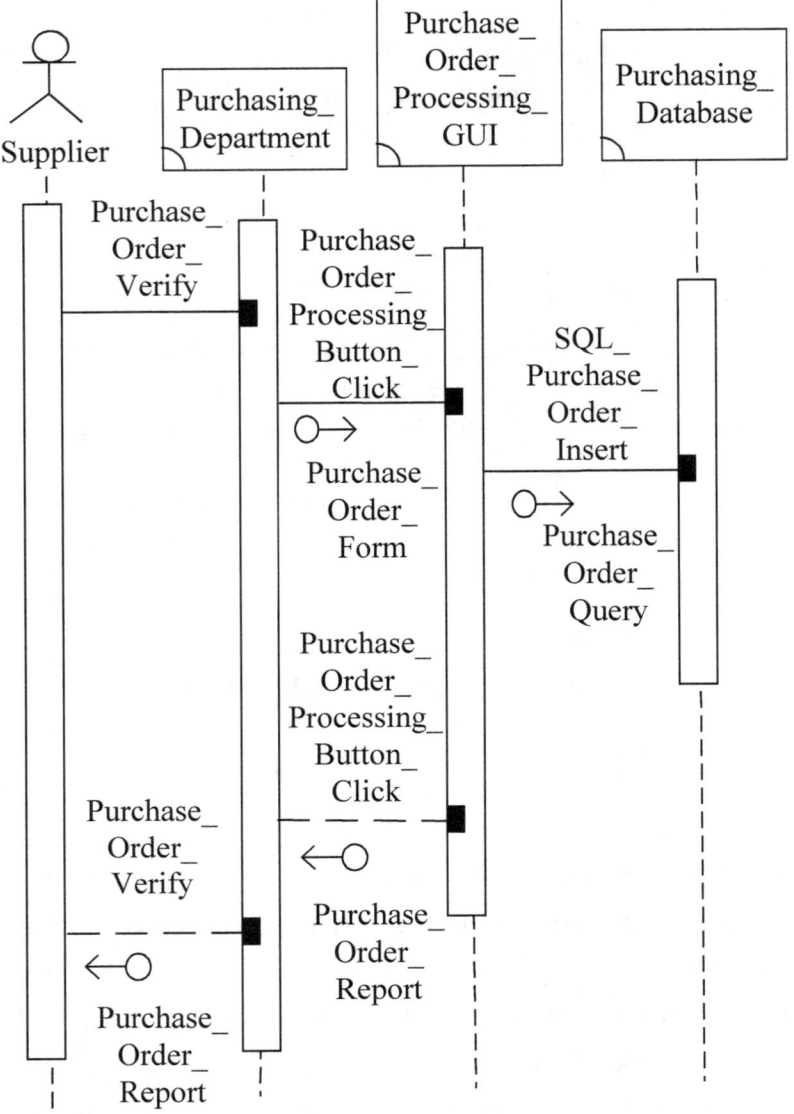

Figure 21-23   IFD of the *Purchase_Order* Behavior

Figure 21-24 shows an IFD of the *Purchase* behavior. First, actor *Supplier* interacts with the *Purchasing_Department* component through the *Purchase_Verify* operation call interaction, carrying the *Products_In* and *Invoice* input parameters. Next, component *Purchasing_Department* interacts with the *Purchase_Processing_GUI* component through the *Purchase_Processing_Button_Click* operation call interaction, carrying the *Purchase_Form* input parameter. Continuingly, component

Purchase_Processing_GUI interacts with the *Purchasing_Database* component through the *SQL_Purchase_Insert* operation call interaction, carrying the *Purchase_Query* input parameter. Finally, actor *Supplier* interacts with the *Purchasing_Department* component through the *Purchase_Verify* operation return interaction, carrying the *Products_Out* output parameter.

Figure 21-24   IFD of the *Purchase* Behavior

Figure 21-25 shows an IFD of the *Collect_Supplier_Data* behavior. First, actor *Supplier* interacts with the *Purchasing_Department* component through the *Interview* operation call interaction. Next, component *Purchasing_Department* interacts with the *Supplier_Data_Processing_GUI* component through the *Supplier_Data_Processing_Button_Click* operation call interaction, carrying the *Supplier_Data_Form* input parameter. Finally, component

214

Supplier_Data_Processing_GUI interacts with the *Purchasing_Database* component through the *SQL_Supplier_Data_Insert* operation call interaction, carrying the *Supplier_Data_Query* input parameter.

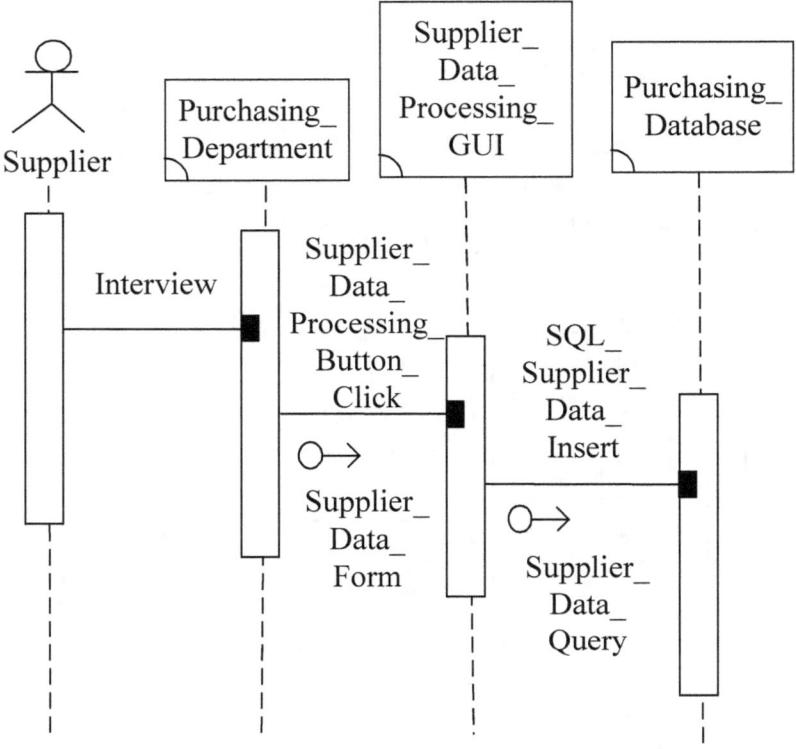

Figure 21-25   IFD of the *Collect_Supplier_Data* Behavior

Chapter 22: Systems Thinking 2.0 Defining the Restaurant

This chapter demonstrates how to achieve systems thinking 2.0 (architectural thinking) defining the *restaurant*, through the application of SBC architecture description language (SBC-ADL).

Supposedly, this restaurant serves Chinese cuisine and French cuisine and Mexican cuisine. The overall behavior of the *restaurant* is represented by two individual behaviors: *Ordering_a_Meal* and *Paying_the_Bill* as shown in Figure 22-1.

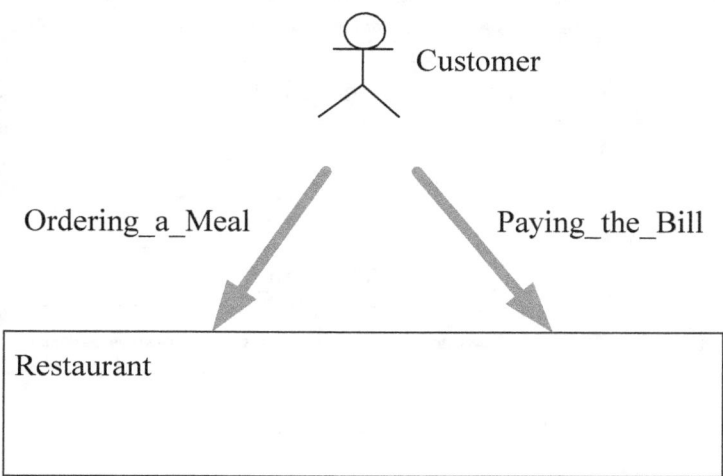

Figure 22-1   Two Behaviors of the *Restaurant*

Using the SBC architecture description language, we shall go through: a) architecture hierarchy diagram, b) framework diagram, c) component operation diagram, d) component connection diagram, e) structure-behavior coalescence diagram and f) interaction flow diagram, to accomplish systems thinking 2.0 (architectural thinking) defining the *restaurant*.

22-1 Architecture Hierarchy Diagram

We use an architecture hierarchy diagram (AHD) to define the multi-level composition and decomposition of the *restaurant*. AHD is the first fundamental diagram to achieve structure-behavior coalescence. As shown in Figure 22-2, *Restaurant* is composed of *Waiting_Staff*, *Cashier* and *Subsystem_1*; *Subsystem_1* is composed of *Chinese_Cuisine_Chef*, *French_Cuisine_Chef* and *Mexican_Cuisine_Chef*.

Figure 22-2   AHD of the *Restaurant*

In Figure 22-2, *Restaurant* and *Subsystem_1* are aggregated systems while *Waiting_Staff*, *Cashier*, *Chinese_Cuisine_Chef*, *French_Cuisine_Chef* and *Mexican_Cuisine_Chef* are non-aggregated systems.

22-2 Framework Diagram

We use a framework diagram (FD) to define the multi-layer composition and decomposition of the *restaurant* as shown in Figure 22-3. FD is the second fundamental diagram to achieve structure-behavior coalescence. In the figure, *Business_SubLayer_2* contains the *Waiting_Staff*, *Cashier* component; *Business_SubLayer_1* contains the *Chinese_Cuisine_Chef*, *French_Cuisine_Chef* and *Mexican_Cuisine_Chef* components.

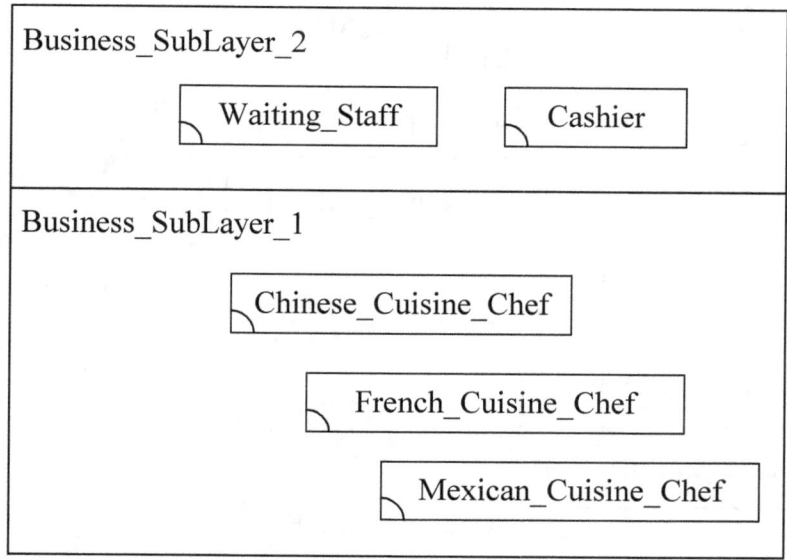

Figure 22-3   FD of the *Restaurant*

22-3 Component Operation Diagram

We use a component operation diagram (COD) to define the operations of all components of the *restaurant* as shown in Figure 22-4. COD is the third fundamental diagram to achieve structure-behavior coalescence. In the figure, component *Waiting_Staff* has one operation: *Take_Order*; *Cashier* has one operation: *Pay_Bills*; component *Chinese_Cuisine_Chef* has one operation: *Cook_Chinese_Food*; component *French_Cuisine_Chef* has one operation: *Cook_French_Food*; component *Mexican_Cuisine_Chef* has one operation: *Cook_Mexican_Food*.

218

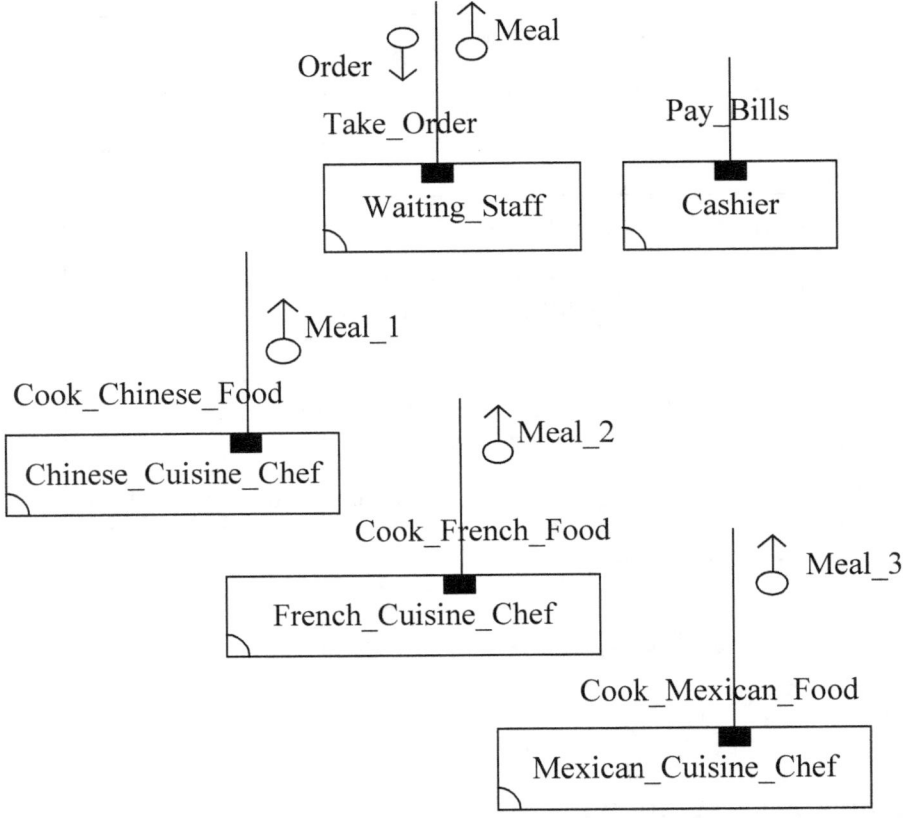

Figure 22-4   COD of the *Restaurant*

The operation formula of *Take_Order* is *Take_Order(In Order; Out Meal)*. The operation formula of *Pay_Bills* is *Pay_Bills*. The operation formula of *Cook_Chinese_Food* is *Cook_Chinese_Food(Out Meal_1)*. The operation formula of *Cook_French_Food* is *Cook_French_Food(Out Meal_2)*. The operation formula of *Cook_Mexican_Food* is *Cook_Mexican_Food(Out Meal_3)*.

Figure 22-5 shows the primitive data type specification of the *Order* input parameter and the *Meal, Meal_1, Meal_2 and Meal_3* output parameters.

Parameter	Data Type	Instances
Order	Enumerated	Chinese, French, Mexican
Meal	Enumerated food	Chinese food, French food, Mexican food
Meal_1	Food	Chinese food
Meal_2	Food	French food
Meal_3	Food	Mexican food

Figure 22-5   Primitive Data Type Specification

22-4 Component Connection Diagram

We use a component connection diagram (CCD) to define the connections among components and actors of the *restaurant* as shown in Figure 22-6. CCD is the fourth fundamental diagram to achieve structure-behavior coalescence.

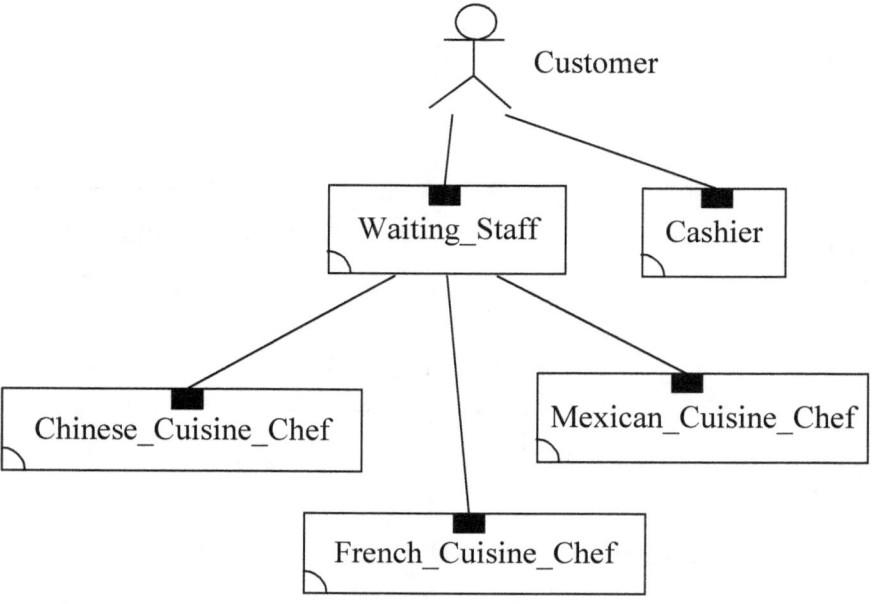

Figure 22-6   CCD of the *Restaurant*

220

In Figure 22-6, actor *Customer* has a connection with each one of the *Waiting_Staff, Cashier* components; component *Waiting_Staff* has a connection with each one of the *Chinese_Cuisine_Chef*, *French_Cuisine_Chef* and *Mexican_Cuisine_Chef* components.

22-5 Structure-Behavior Coalescence Diagram

We use a structure-behavior coalescence diagram (SBCD) to define the systems structure and systems behavior coexisting in the *restaurant* as shown in Figure 22-7. SBCD is the fifth fundamental diagram to achieve structure-behavior coalescence. In the figure, interactions among the *Customer* actor and the *Waiting_Staff*, *Cashier*, *Chinese_Cuisine_Chef*, *French_Cuisine_Chef*, *Mexican_Cuisine_Chef* components shall draw forth the *Ordering_a_Meal* and *Paying_the_Bill* behaviors.

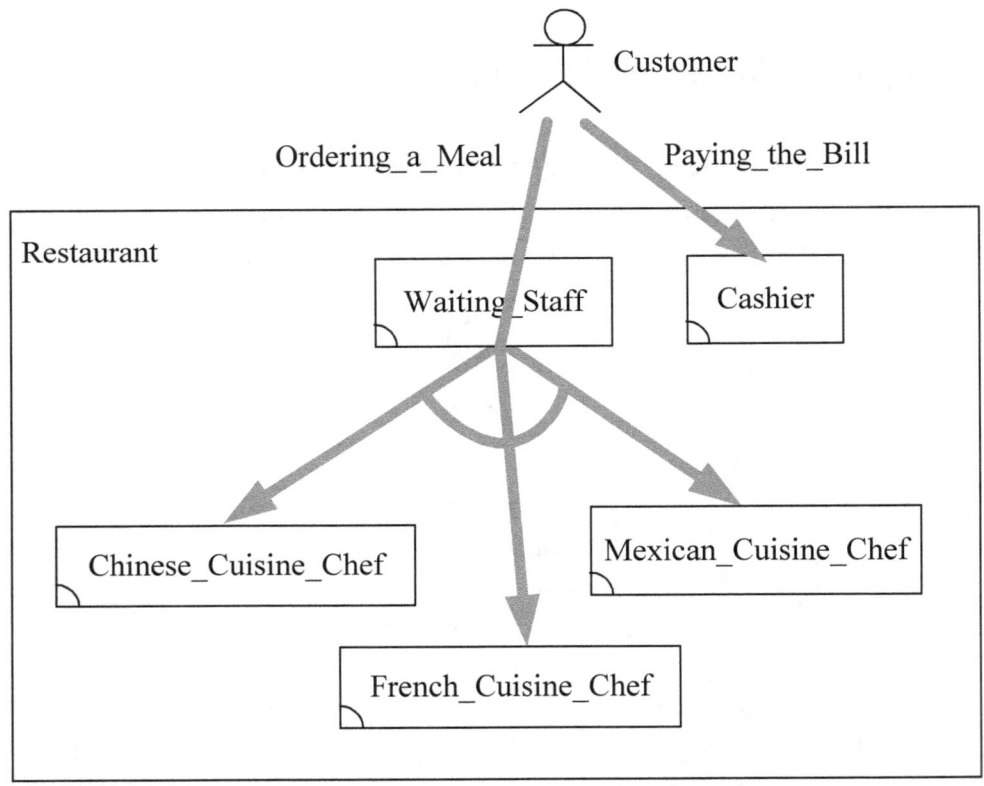

Figure 22-7   SBCD of the *Restaurant*

The overall behavior of a system is the aggregation of all its individual behaviors. For example, the overall behavior of the *restaurant* includes the *Ordering_a_Meal* and *Paying_the_Bill* behaviors.

The major purpose of SBC architecture description language is to define an integrated whole of a system. In Figure 22-7, we not only see its structure, but also see at the same time its behavior in the *restaurant*'s SBCD.

22-6 Interaction Flow Diagram

The overall behavior of the *restaurant* includes two individual behaviors: *Ordering_a_Meal* and *Paying_the_Bill*. Each individual behavior is represented by an execution path. We use an interaction flow diagram (IFD) to define each one of these execution paths. IFD is the sixth fundamental diagram to achieve structure-behavior coalescence.

Figure 22-8 shows an IFD of the *Ordering_a_Meal* behavior. First, actor *Customer* interacts with the *Waiting_Staff* component through the *Take_Order* operation call interaction, carrying the *Order* input parameter. Next, if the *Order = "Chinese"* condition is true then component *Waiting_Staff* shall interact with the *Chinese_Cuisine_Chef* component through the *Cook_Chinese_Food* conditional operation call interaction; else if the *Order = "French"* condition is true then component *Waiting_Staff* shall interact with the *French_Cuisine_Chef* component through the *Cook_French_Food* conditional operation call interaction; else component *Waiting_Staff* shall interact with the *Mexican_Cuisine_Chef* component through the *Cook_Mexican_Food* conditional operation call interaction. Continuingly, if the *Order = "Chinese"* condition is true then component *Waiting_Staff* shall interact with the *Chinese_Cuisine_Chef* component through the *Cook_Chinese_Food* conditional operation return interaction, carrying the *Meal_1* output parameter; else if the *Order = "French"* condition is true then component *Waiting_Staff* shall interact with the *French_Cuisine_Chef* component through the *Cook_French_Food* conditional operation return interaction, carrying the *Meal_2* output parameter; else component *Waiting_Staff* shall interact with the *Mexican_Cuisine_Chef* component through the *Cook_Mexican_Food* conditional operation return interaction, carrying the *Meal_3* output parameter. Finally, actor *Customer* interacts with the *Waiting_Staff* component through the *Take_Order* operation return interaction, carrying the *Meal* output parameter.

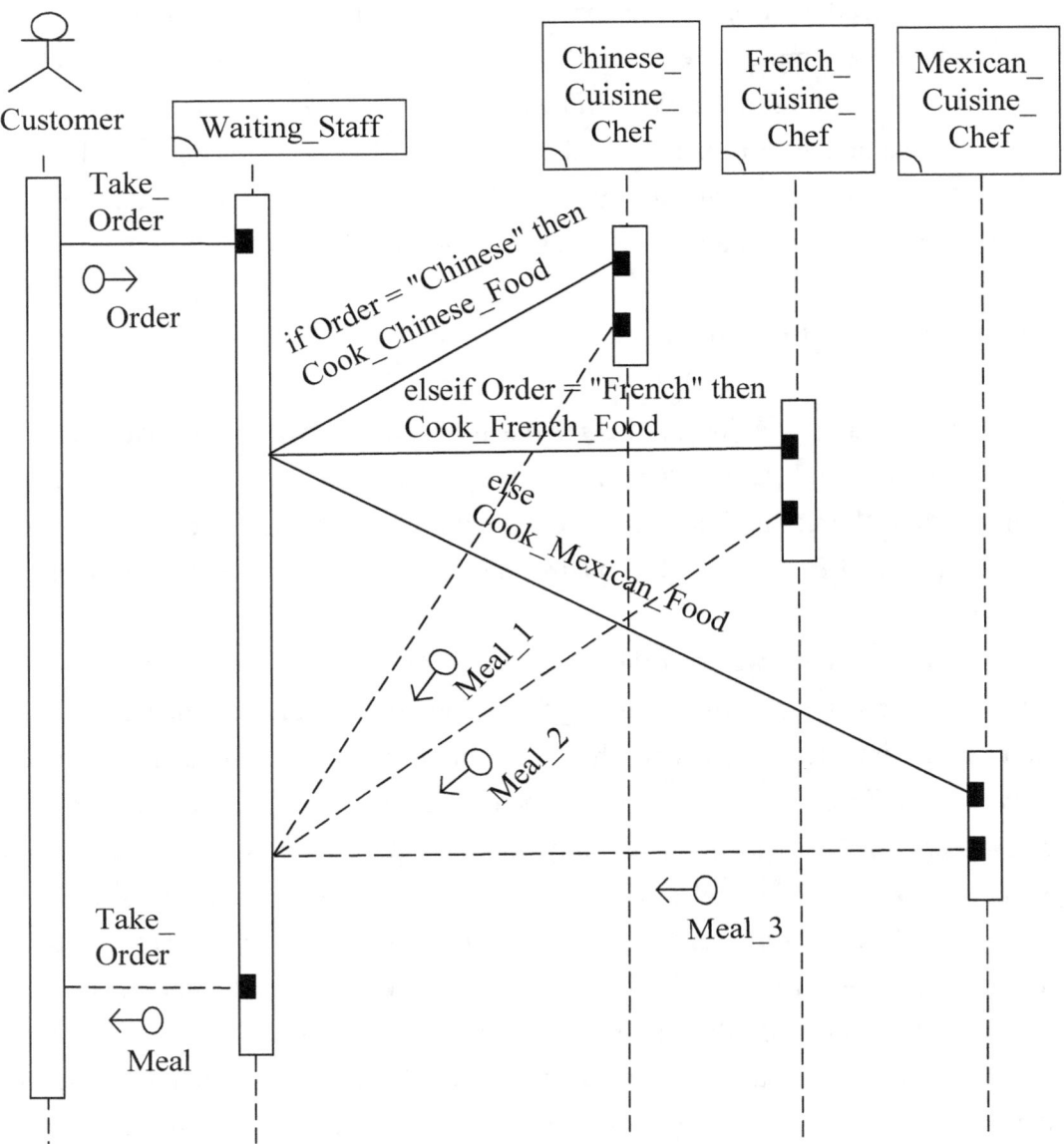

Figure 22-8   IFD of the *Ordering_a_Meal* Behavior

Figure 22-9 shows an IFD of the *Paying_the_Bill* behavior. First, actor *Customer* interacts with the *Cashier* component through the *Pay_Bills* operation call interaction.

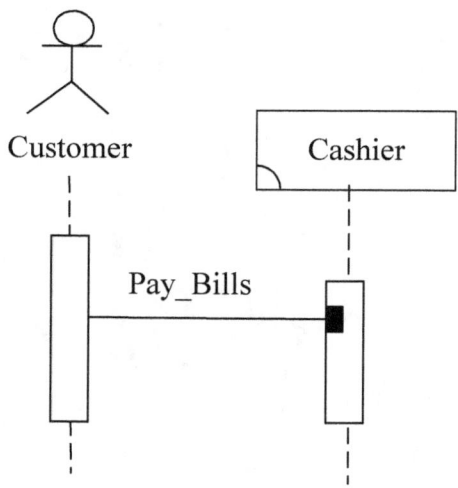

Figure 22-9   IFD of the *Paying_the_Bill* Behavior

Chapter 23: Systems Thinking 2.0 Defining the Car

This chapter demonstrates how to achieve systems thinking 2.0 (architectural thinking) defining the *car*, through the application of SBC architecture description language (SBC-ADL).

Generally, the overall behavior of the *car* is prominently represented by three individual behaviors: *Setting_Gear_to_Forward*, *Setting_Gear_to_Reverse* and *Driving* as shown in Figure 23-1.

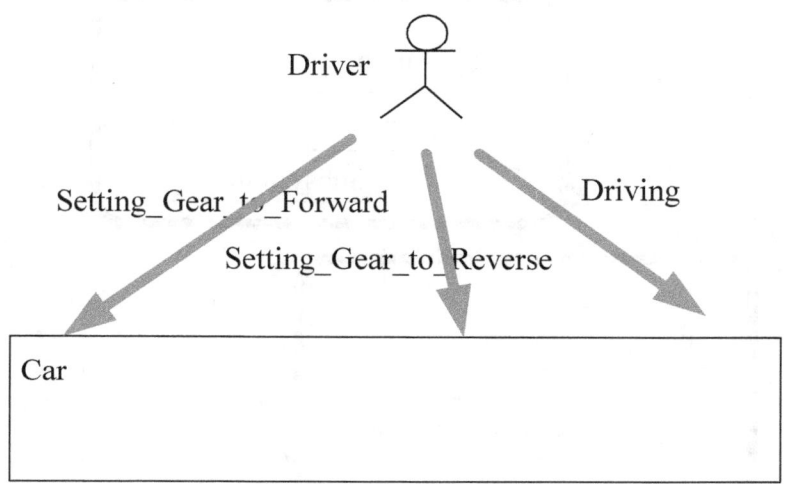

Figure 23-1   Three Behaviors of the *Car*

Using the SBC architecture description language, we shall go through: a) architecture hierarchy diagram, b) framework diagram, c) component operation diagram, d) component connection diagram, e) structure-behavior coalescence diagram and f) interaction flow diagram, to accomplish systems thinking 2.0 (architectural thinking) defining the *car*.

23-1 Architecture Hierarchy Diagram

We use an architecture hierarchy diagram (AHD) to define the multi-level composition and decomposition of the *car*. AHD is the first fundamental diagram to achieve structure-behavior coalescence. As shown in Figure 23-2, *Car* is composed of *Transmission*, *Gas_Pedal* and *Subsystem_2*; *Subsystem_2* is composed of *Engine* and *Subsystem_1*; *Subsystem_1* is composed of *Forward_Gear* and *Reverse_Gear*.

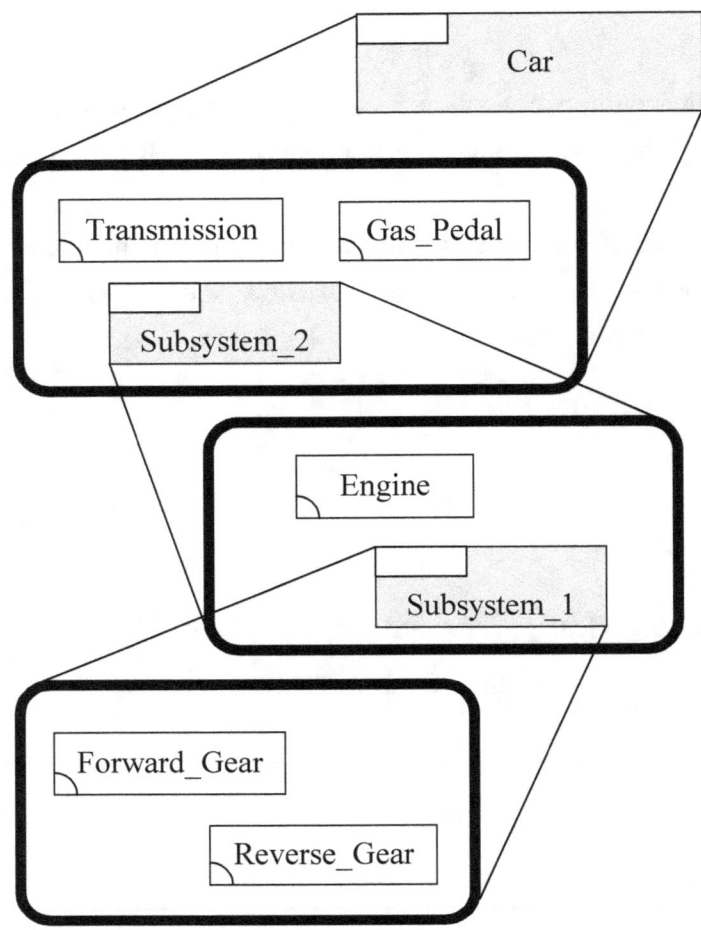

Figure 23-2   AHD of the *Car*

In Figure 20-2, *Car*, *Subsystem_2* and *Subsystem_1* are aggregated systems while *Transmission*, *Gas_Pedal*, *Engine*, *Forward_Gear and Reverse_Gear* are non-aggregated systems.

23-2 Framework Diagram

We use a framework diagram (FD) to define the multi-layer composition and decomposition of the *car* as shown in Figure 23-3. FD is the second fundamental diagram to achieve structure-behavior coalescence. In the figure, *Technology_SubLayer_3* contains the *Transmission* and *Gas_Pedal* components; *Technology_SubLayer_2* contains the *Engine* component; *Technology_SubLayer_1* contains the *Forward_Gear* and *Reverse_Gear* components.

Figure 23-3   FD of the *Car*

23-3 Component Operation Diagram

We use a component operation diagram (COD) to define the operations of all components of the *car* as shown in Figure 23-4. COD is the third fundamental diagram to achieve structure-behavior coalescence. In the figure, component *Transmission* has two operations: *Push_Gear_Forward* and *Push_Gear_Reverse*; component *Gas_Pedal* has one operation: *Depress_Gas_Pedal*; component *Engine* has one operation: *Fuel_Supply*; component *Forward_Gear* has one operation: *Forward_Gear_Rotate*; component *Reverse_Gear* has one operation: *Reverse_Gear_Rotate*.

Figure 23-4   COD of the *Car*

23-4 Component Connection Diagram

We use a component connection diagram (CCD) to define the connections among components and actors of the *car* as shown in Figure 23-5. CCD is the fourth fundamental diagram to achieve structure-behavior coalescence.

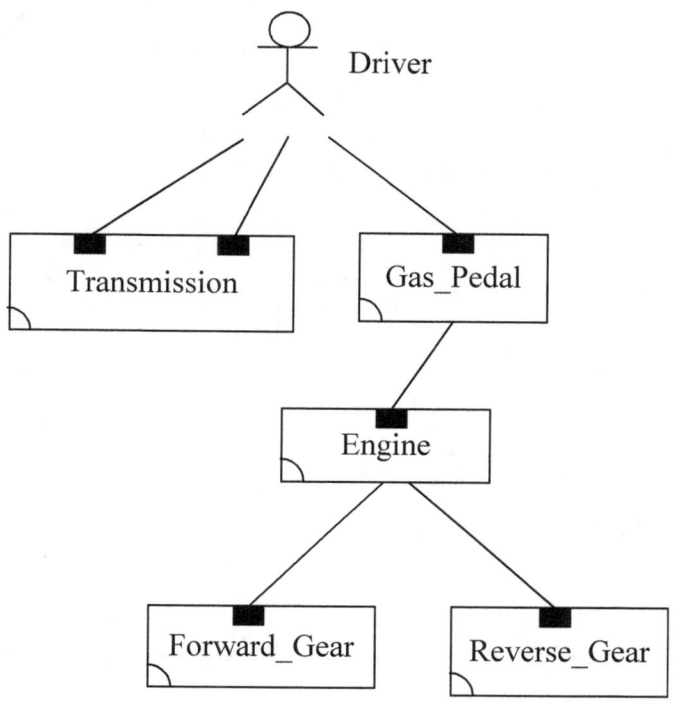

Figure 23-5   CCD of the *Car*

In Figure 23-5, actor *Driver* has two connections with the *Transmission* component; actor *Driver* has a connection with the *Gas_Pedal* component; component *Gas_Pedal* has a connection with the *Engine* component; component *Engine* has a connection with one of the *Forward_Gear* and *Reverse_Gear* components.

23-5 Structure-Behavior Coalescence Diagram

We use a structure-behavior coalescence diagram (SBCD) to define the systems structure and systems behavior coexisting in the *car* as shown in Figure 23-6. SBCD is the fifth fundamental diagram to achieve structure-behavior coalescence. In the figure, interactions among the *Driver* actor and the *Transmission*, *Gas_Pedal*, *Engine*, *Forward_Gear*, *Reverse_Gear* components shall draw forth the *Setting_Gear_to_Forward*, *Setting_Gear_to_Reverse* and *Driving* behaviors.

230

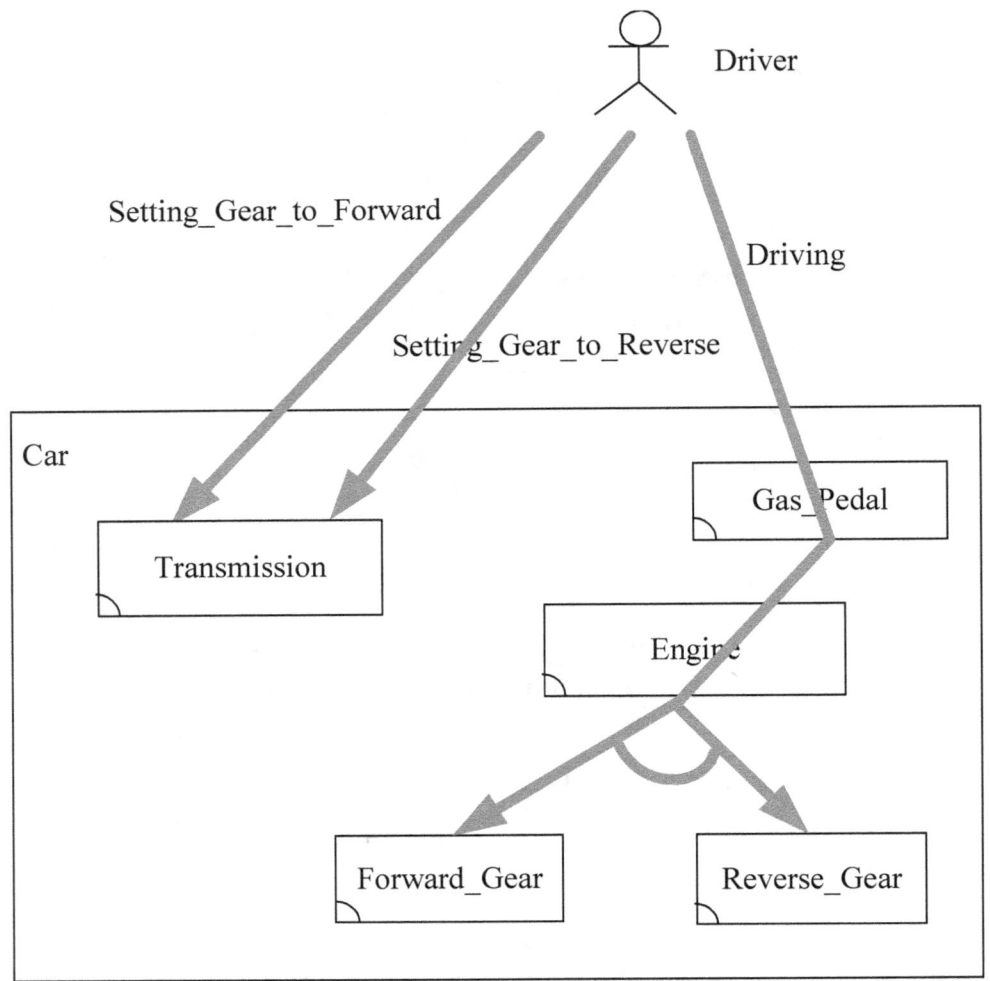

Figure 23-6   SBCD of the *Car*

The overall behavior of a system is the aggregation of all its individual behaviors. For example, the overall behavior of the *car* includes the *Setting_Gear_to_Forward*, *Setting_Gear_to_Reverse* and *Driving* behaviors.

Be noticed that the *Setting_Gear_to_Forward*, *Setting_Gear_to_Reverse* and *Driving* behaviors are mutually independent of each other. They tend to be executed concurrently [Hoar85, Miln89, Miln99].

The major purpose of SBC architecture description language is to define an integrated whole of a system. In Figure 23-6, we not only see its structure, but also see at the same time its behavior in the *car*'s SBCD.

23-6 Interaction Flow Diagram

The overall behavior of the *car* includes three individual behaviors: *Setting_Gear_to_Forward*, *Setting_Gear_to_Reverse* and *Driving*. Each individual behavior is represented by an execution path. We use an interaction flow diagram (IFD) to define each one of these execution paths. IFD is the sixth fundamental diagram to achieve structure-behavior coalescence.

Figure 23-7 shows an IFD of the *Setting_Gear_to_Forward* behavior. First, actor *Driver* interacts with the *Transmission* component through the *Push_Gear_Forward* operation call interaction, assigning the "Forward" value to the *Gear* global variable.

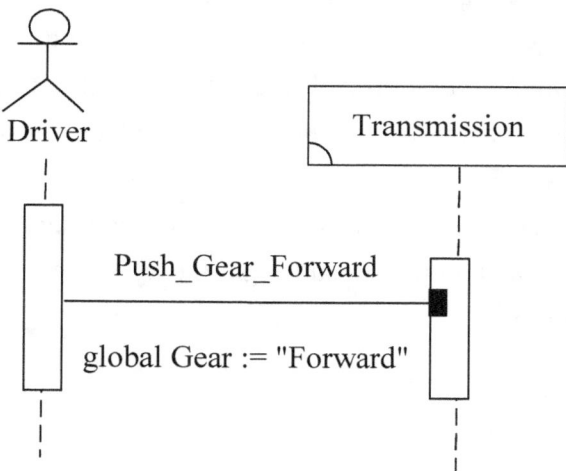

Figure 23-7   IFD of the *Setting_Gear_to_Forward* Behavior

Figure 23-8 shows an IFD of the *Setting_Gear_to_Reverse* behavior. First, actor *Driver* interacts with the *Transmission* component through the *Push_Gear_Reverse* operation call interaction, assigning the "*Reverse*" value to the *Gear* global variable.

232

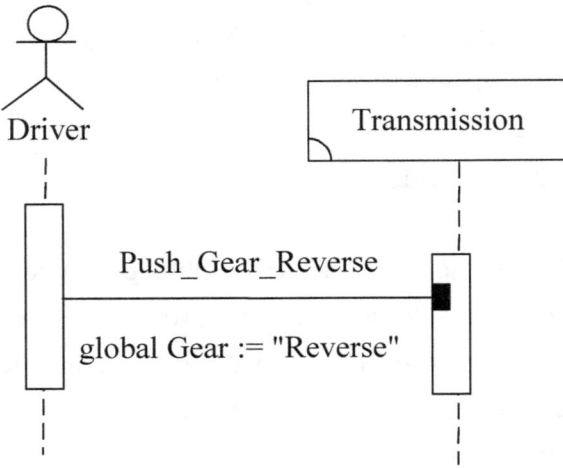

Figure 23-8   IFD of the *Setting_Gear_to_Reverse* Behavior

Figure 23-9 shows an IFD of the *Driving* behavior. First, actor *Driver* interacts with the *Gas_Pedal* component through the *Depress_Gas_Pedal* operation call interaction. Second, component *Gas_Pedal* interacts with the *Engine* component through the *Fuel_Supply* operation call interaction. Finally, if the *Gear = "Forward"* condition is true then component *Engine* shall interact with the *Forward_Gear* component through the *Forward_Gear_Rotate* conditional operation call interaction; else if the *Gear = "Reverse"* condition is true then component *Engine* shall interact with the *Reverse_Gear* component through the *Reverse_Gear_Rotate* conditional operation call interaction.

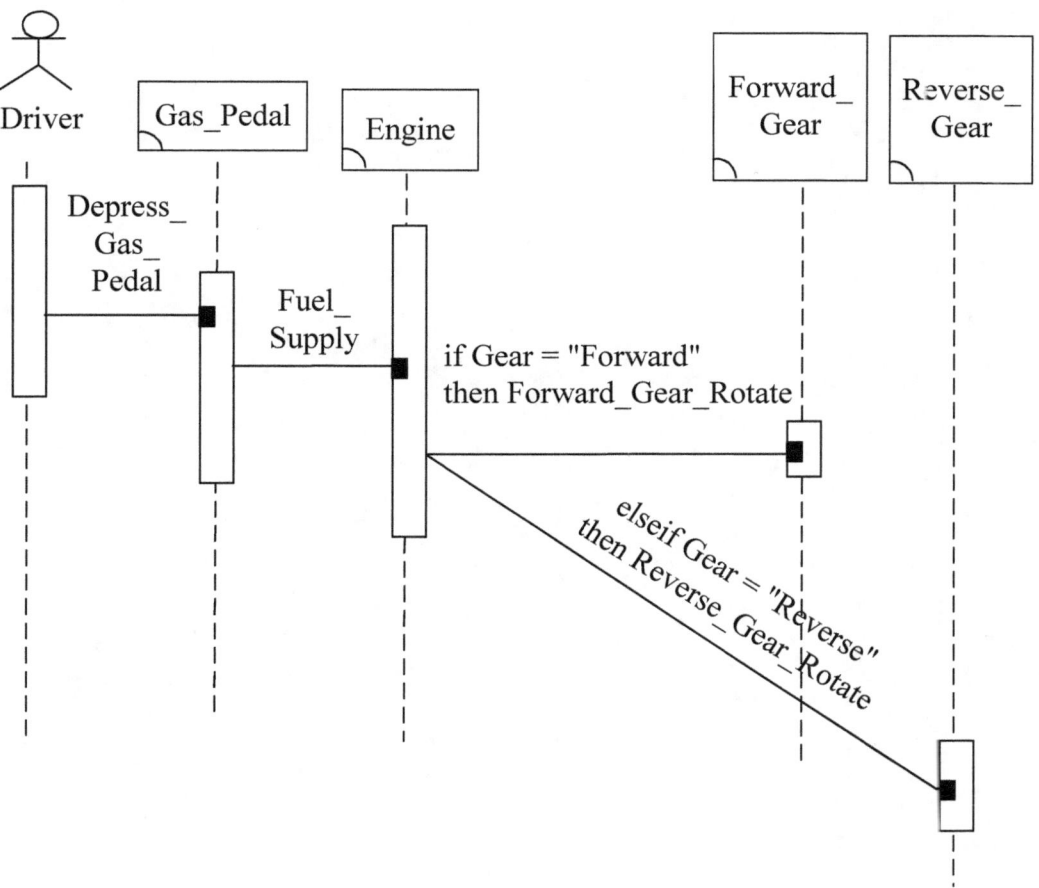

Figure 23-9   IFD of the *Driving* Behavior

APPENDIX: SBC ARCHITECTURE DESCRIPTION LANGUAGE

(1) Architecture Hierarchy Diagram

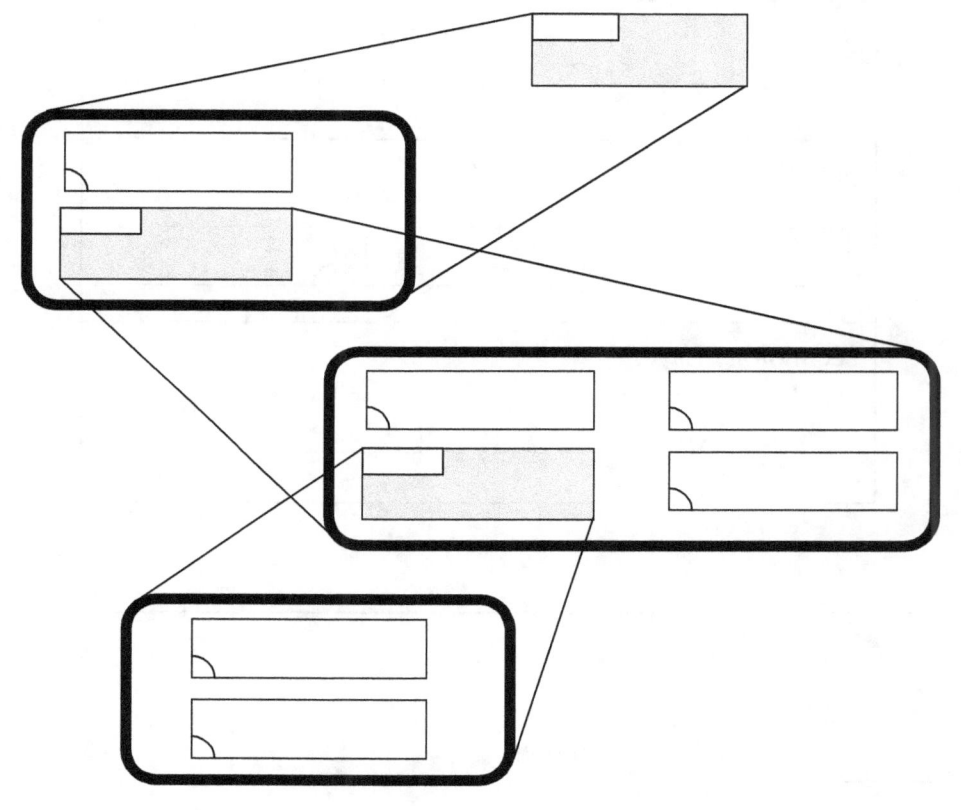

: Aggregated System

: Non-Aggregated System,   Component

(2) Framework Diagram

Business Layer

Application Layer

Data Layer

Technology Layer

: Component

(3) Component Operation Diagram

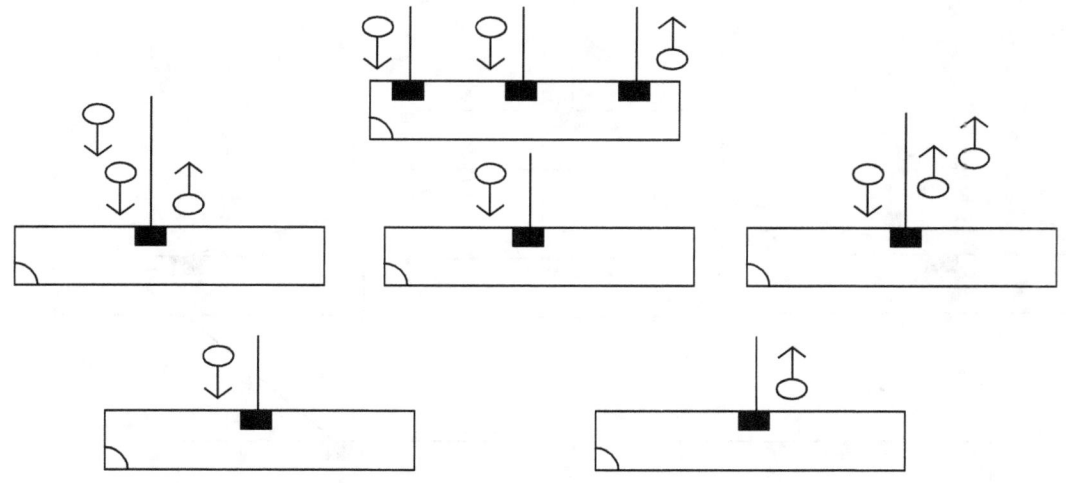

■ : Operation

↓○ : Input Data

↑○ : Output Data

▭ : Component

(4) Component Connection Diagram

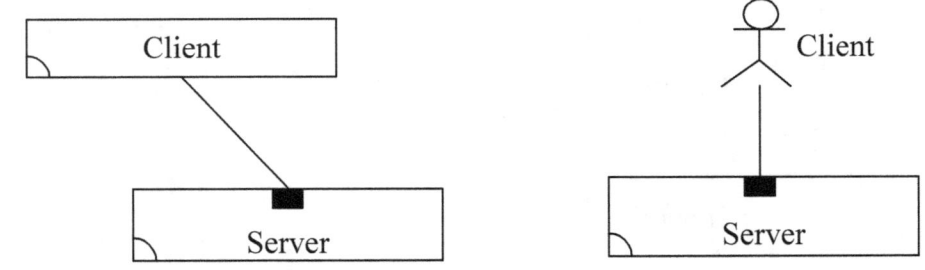

(5) Structure-Behavior Coalescence Diagram

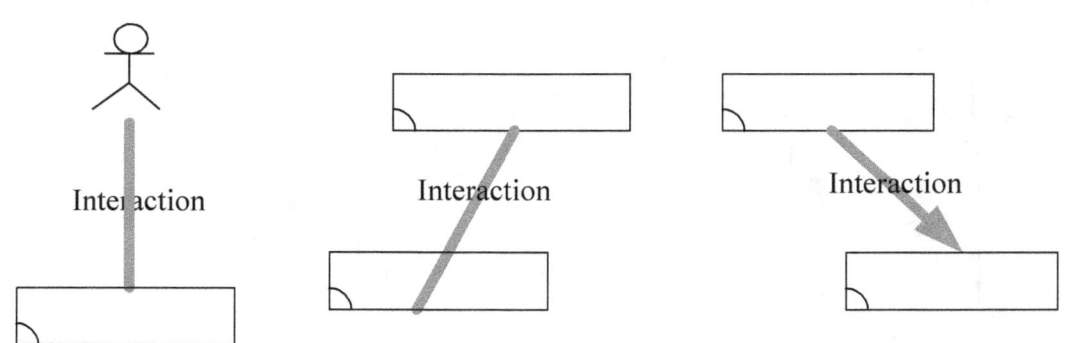

(6) Interaction Flow Diagram

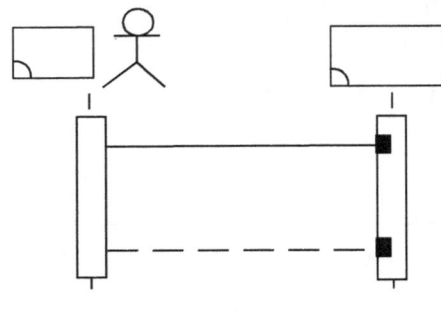

: Operation Call Interaction

: Operation Return Interaction

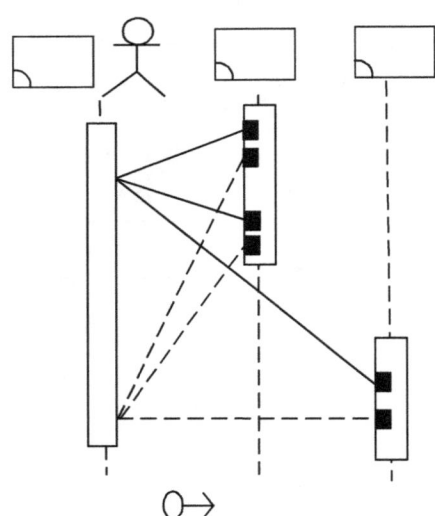

: Conditional
Operation Call Interaction

: Conditional
Operation Return Interaction

O→ : Input Data

←O : Output Data

BIBLIOGRAPHY

[Acko68] Ackoff, R., "Toward a System of Systems Concepts," *Modern Systems Research for the Behavioral Scientist: A Sourcebook*, Aldine Publishing Company, 1968.

[Bare84] Barendregt, H. P., *The Lambda Calculus: Its Syntax and Semantics*, Elsevier Science Publishers, 1984.

[Beam90] Beam, W. R., *Systems Engineering: Architecture and Design*, McGraw-Hill, 1990.

[Bere09] Berenbach, B. et al., *Software & Systems Requirements Engineering: In Practice*, McGraw-Hill Osborne Media, 1st Edition, 2009.

[Bert69] Von Bertalanffy, L., *General System Theory: Foundations, Development, Applications*, George Braziller Inc., Revised Edition, 1969.

[Bert81] Von Bertalanffy, L. et al., *Systems View of Man: Collected Essays*, Westview Pr, 1981.

[Burd10] Burd, S. D., *Systems Architecture*, 6th Edition, Cengage Learning, 2010.

[Chao09] Chao, W. S. et al., *System Analysis and Design: SBC Software Architecture in Practice*, LAP Lambert Academic Publishing, 2009.

[Chao12] Chao, W. S., *Systems Architecture: SBC Architecture at Work*, LAP Lambert Academic Publishing, 2012.

[Chao14] Chao, W. S., *General Systems Theory 2.0: General Architectural Theory Using the SBC Architecture*, CreateSpace Independent Publishing Platform, 2014.

[Chec99] Checkland, P., *Systems Thinking, Systems Practice: Includes a 30-Year Retrospective*, Wiley, 1st Edition, 1999.

[Date03] Date, C. J., *An Introduction to Database Systems*, 8th Edition, Addison Wesley, 2003.

[Elma10] Elmasri, R., *Fundamentals of Database Systems*, 6th Edition, Addison Wesley, 2010.

[Frie11] Friedenthal, S., et al., *A Practical Guide to SysML: The Systems Modeling Language*, Morgan Kaufmann, 2nd Edition, 2011.

[Gali07] Galitz, W., *The Essential Guide to User Interface Design: An Introduction to GUI Design Principles and Techniques*, 3rd Edition Wiley, 2007.

[Gall03] Gall, J., *The Systems Bible: The Beginner's Guide to Systems Large and Small*, General Systemantics Pr/Liberty, 2003.

[Ghar11] Gharajedaghi, J., *Systems Thinking: Managing Chaos and Complexity: A Platform for Designing Business Architecture*, Morgan Kaufmann, 2011.

[Grad06] Grady, J. O., *System Requirements Analysis*, Academic Press, 1st Edition, 2006.

[Hend80] Henderson, P., *Functional Programming: Application and Implementation*, Prentice-Hall, 1980.

[Hoar85] Hoare, C. A. R., *Communicating Sequential Processes*, Prentice-Hall, 1985.

[Hoff10] Hoffer, J. A., et al., *Modern Systems Analysis and Design*, Prentice Hall, 6th Edition, 2010.

[Jorg12] Jorgensen, S. E., *Introduction to Systems Ecology (Applied Ecology and Environmental Management)*, CRC Press, 2012.

[Kapo94] Kaposi, A., et al., *Systems, Models and Measure*, Springer-Verlag London Limited, 1994.

[Kass07] Kasser, J. E., *A Framework for Understanding Systems Engineering*, BookSurge Publishing, 2007.

[Kill09] Killoran, D. M., *LSAT Logical Reasoning Bible: A Comprehensive System for Attacking the Logical Reasoning Section of the LSAT*, PowerScore Publishing, 2009.

[Klip09] Klipp, E. et al., *Systems Biology: A Textbook*, 1st Edition, Wiley-VCH, 2009.

[Koss11] Kossiakoff, A. et al., *Systems Engineering Principles and Practice*, 2nd Edition, Wiley-Interscience, 2011.

[Lasz96] Laszlo, E., *The Systems View of the World: A Holistic Vision for Our Time*, 2nd Edition, Hampton Pr, 1996.

[Luhm12] Luhmann, N., *Introduction to Systems Theory*, 1st Edition, Polity, 2012.

[Mann74] Manna, Z., *Mathematical Theory of Computation*, McGraw-Hill, 1974.

[Maie09] Maier, M. W., *The Art of Systems Architecting*, 3rd Edition, CRC Press, 2009.

[Mead08] Meadows, D. H., *Thinking in Systems: A Primer*, Chelsea Green Publishing, 2008.

[Miln89] Milner, R., *Communication and Concurrency*, Prentice-Hall, 1989.

[Miln99] Milner, R., *Communicating and Mobile Systems: the π-Calculus*, 1st Edition, Cambridge University Press, 1999.

[Mull11] Muller, G., *Systems Architecting: A Business Perspective*, CRC Press, 2011.

[Odum94] Odum, H. T., *Ecological and General Systems: An Introduction to Systems Ecology*, Rev Sub Edition, University Press of Colorado, 1994.

[Ogat03] Ogata, K., *System Dynamics*, 4th Edition, Prentice Hall, 2003.

[Palm09] Palm, W. III, *System Dynamics*, 2nd Edition, McGraw-Hill Science/Engineering/Math, 2009.

[Pork78] Porkert, M., *Theoretical Foundations of Chinese Medicine: Systems of Correspondence*, The MIT Press, 1978.

[Prat00] Pratt, T. W. et al., *Programming Languages: Design and Implementation*, 4th Edition, Prentice Hall 2000.

[Pres09] Pressman, R. S., *Software Engineering: A Practitioner's Approach*, 7th Edition, McGraw-Hill, 2009.

[Raff11] Raff, H. et al., *Medical Physiology: A Systems Approach*, 1st Edition, McGraw-Hill Professional, 2011.

[Roza11] Rozanski, N. et al., *Software Systems Architecture: Working With Stakeholders Using Viewpoints and Perspectives*, 2nd Edition, Addison-Wesley Professional, 2011.

[Sang03] Sangiorgi, D. et al., *The Pi-Calculus: A Theory of Mobile Processes*, Cambridge University Press, 2003.

[Scho10] Scholl, C., *Functional Decomposition with Applications to FPGA Synthesis*, Springer, 2010.

[Sebe12] Sebesta, R. W., *Programming the World Wide Web*, 7th Edition, Addison-Wesley, 2012.

[Seth96] Sethi, R., *Programming Languages: Concepts and Constructs*, 2nd Edition, Addison-Wesley, 1996.

[Shap00] Shapiro. S., *Foundations without Foundationalism: A Case for Second-order Logic*, Oxford University Press, 2000.

[Shel11] Shelly, G. B., et al., *Systems Analysis and Design*, 9th Edition, Course Technology, 2011.

[Sher09] Sherwood, L., *Human Physiology: From Cells to Systems*, 7th Edition, Brooks Cole, 2009.

[Somm06] Sommerville, I., *Software Engineering*, 8th Edition, Addison-Wesley, 2006.

[Voit12] Voit, E., *A First Course in Systems Biology*, 1st Edition, Garland Science, 2012.

[Wall04] Wall, D., *Multi-Tier Application Programming with PHP: Practical Guide for Architects and Programmers*, Morgan Kaufmann, 2004.

[Warf06] Warfield, J. N., *An Introduction to Systems Science*, World Scientific Publishing Company, 2006.

[Weil00] Weil, A., *Spontaneous Healing: How to Discover and Embrace Your Body's Natural Ability to Maintain and Heal Itself*, Ballantine Books, 2000.

[Weil04] Weil, A., *Health and Healing: The Philosophy of Integrative Medicine and Optimum Health*, Revised Edition, Mariner Books, 2004.

INDEX

T